A Quick Example Concerning Expectations

Discussion of a math problem:

Suppose we have an 8×8 array of squares, and a supply of 2×1 tiles capable of and allowed only to cover exactly two contiguous squares.

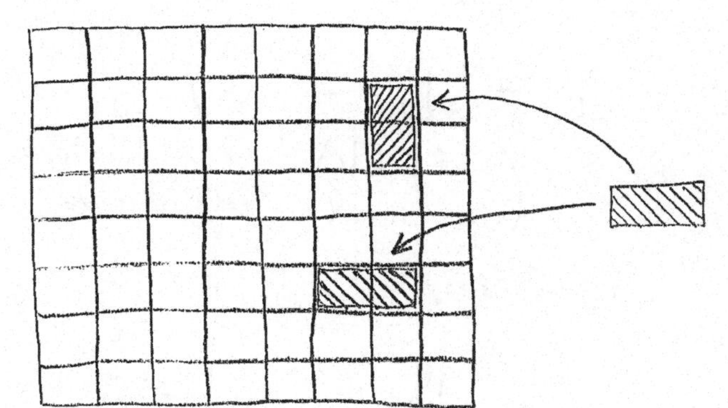

Clearly 32 of these tiles can be used to cover the 64 squares.

But suppose we remove 2 opposite corners:

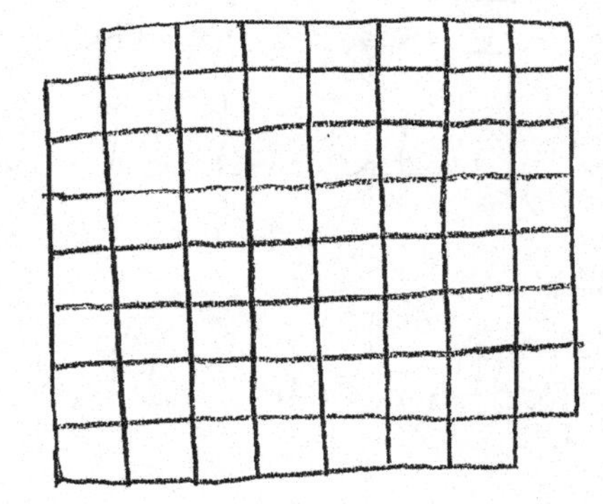

Can 31 tiles cover this modified array?

Trial and error would be difficult to implement, and require billions of trials...

Creative solution

Apply a "checker board" pattern to the array, and note that every tile covers two squares of different color.

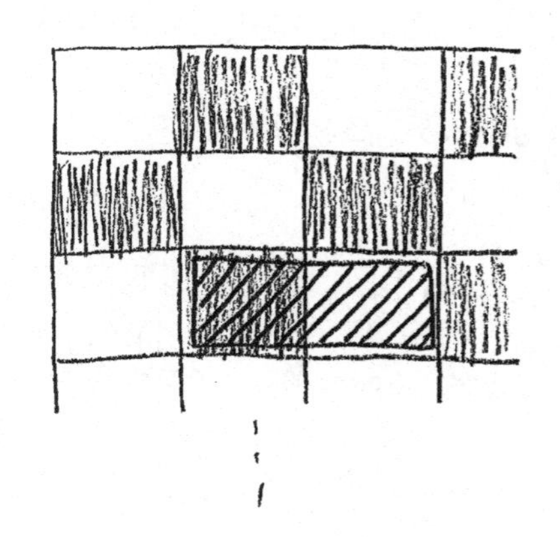

So for every arrangement of tiles,

$$I = B - W$$

(# of black sqs. covered) (# of white sqs. covered)

is zero.

The two removed squares must be the same color.

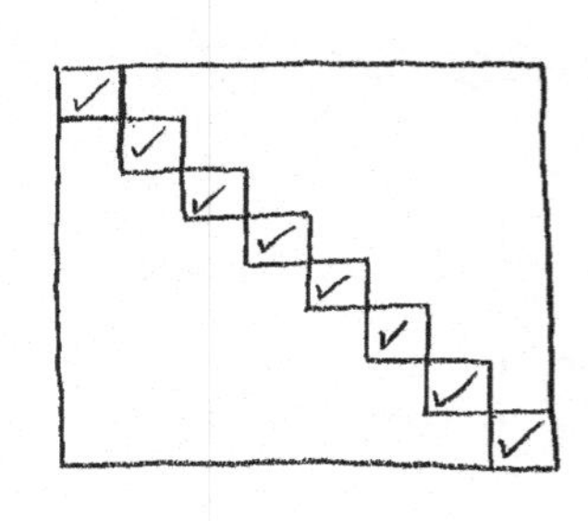

So the desired arrangement would have I equal either 2 or -2, neither of which is possible.

- "Creative", because neither the pattern, nor "I", was given in the question.
- "I" is an example of a powerful mathematical strategy called an "invariant".

Questions for students:

1. 8×8, opposite corners removed; possible?

2. 20×20? 750×750?

3. 10×20? 20×40?

4. 75×150?

5. 100×100, non-opposite corners removed?

6. 100×100, two unspecified squares of different color removed?

7. 300×300, tiles now 3×1, remove any 3 corners.

Comments

- 1. is merely regurgitation.
- 2., 3. apply same idea to new contexts, but modifications to the argument are easy. Must check details!
- 4., 5., 6., argument fails. Also, answer changes, and a completely different argument is needed! (6. is hard!)
- 7. requires creation of new invariants (3 colors!) and new challenges arise...

Applications

We all love applications!

In linear algebra, there are big applications of big ideas. :)

But it takes a long time to develop the big ideas... so the applications come more slowly. :(

High school: math idea, appl., math idea, appl. ......

Linear algebra: math ideas .... application ....

Don't wait for the applications before you invest yourself in the math!

You must buy in to the (well established) idea that the math will be useful!

Some fundamentals of approach in math

Deductive reasoning: Assumptions $\Longrightarrow$ Conclusions

If the assumptions are true, then the conclusions must also be true.

But if the assumptions are not true, the conclusions are not valid!

In math:

Assumptions		Conclusions
Axioms Definitions Previous Theorems Conditions	Proof $\Longrightarrow$	Result

(Conditions $\Longrightarrow$ Result: Theorem)

Mathematics is extremely careful with axioms.

Definitions are very carefully chosen, and adhered to strictly.

Previously established theorems are used thoughtfully.

Proofs must be extremely careful and rigorous.

Def:) A number r is "rational" if it is the ratio $r = m/n$ of two integers m, n with no common factors.

Thm:) If $p > 1$ is prime, then $\sqrt{p}$ is not rational.

Pf:) Suppose $\sqrt{p}$ were rational; then $\sqrt{p} = m/n$ where m, n have no common factors.

Then $p = m^2/n^2$, and so $m^2 = p n^2$.

If p is not a factor of m, then it is also not a factor of m^2; this would be a contradiction, since p is a factor of pn^2.

So $m = pq$, and then $p^2 q^2 = p n^2$, so $p q^2 = n^2$.

But then p must also be a factor n by an analogous argument... in which case we see m, n have a common factor of p, which is a contradiction.

So our original assumption must be false, and $\sqrt{p}$ is not rational.

Misconceptions

1. "Definitions and theorems are roughly the same things."

No! Definitions are chosen; you cannot prove a definition.

Two equivalent statements ($P \Leftrightarrow Q$) describing the same thing could be viewed as "equivalent definitions"; but choose one as the definition.

Ex: $\left(e = \lim_{n\to\infty}\left(1+\frac{1}{n}\right)^n\right) \Leftrightarrow \left(e \text{ is the unique number with } \lim_{h\to 0}\frac{e^h-1}{h} = 1\right)$

2. "Conditions don't matter." or "Conditions are always true."

No! You must check the conditions before applying a theorem, else your result is invalid!

Ex: Theorem: Every even integer is the sum of two odd integers.

Note the condition is that you start with an even integer.

If you apply to an odd integer, or a fraction, the result you obtain is invalid (also false in this case.)

3. "You can prove scientific statements about the physical world."

Not in the mathematical sense.

The relationship between theory and experiment is this:

(*If* a scientific theory is true ...) ⟹ (... *then* the experiment will work as predicted)

The implication only points one way...

What you can prove:

(*If* the experiment does *not* work as predicted...) ⟹ (... *then* the theory is *false*.)

A theory can be viewed as "stronger" by surviving more and more experimental challenges, and in practice this is sufficient for most purposes. But it is *never proved* in the deductive sense.

Ex:) Newton's theories of motion, held for 250 years, but....

Mathematical modeling makes assumptions about the physical world, and then <u>within that model</u> you can deductively prove results:

$$\begin{pmatrix}\text{Assumptions about} \\ \text{physical world...}\end{pmatrix} \Rightarrow \begin{pmatrix}\text{... Conclusions of} \\ \text{the math model}\end{pmatrix}$$

But even experimentally "verifying" the results <u>does not prove</u> the assumptions; and if the assumptions fail, so do the results.

<u>In this course:</u>

1. Students will be expected to adhere to deductive reasoning, to a degree <u>appropriate to this course</u>. (See from context; notes, book, old exams, ...)
2. More than Math 212; less than Math 431...

1.1 – Systems of Linear Equations

How do we solve a system such as:

$$\begin{aligned} a_{11}x_1 + a_{12}x_2 + \cdots + a_{1n}x_n &= b_1 \\ a_{21}x_1 + a_{22}x_2 + \cdots + a_{2n}x_n &= b_2 \\ &\vdots \\ a_{m1}x_1 + a_{m2}x_2 + \cdots + a_{mn}x_n &= b_m \end{aligned}$$

Do we know a solution exists? Might there be more than one solution? How many?

Bad Method: Start deriving more and more new equations...

Ex: Solve

$$\begin{aligned} x - y \quad\;\; &= 1 \\ y - z &= 1 \\ x \quad\;\; - z &= 1 \end{aligned}$$

Attempt:

$$\begin{aligned} 2x - y - z &= 2 \\ x + y - 2z &= 2 \\ 3x \quad\;\; - 3z &= 4 \\ 4x + y - 5z &= 6 \\ &\vdots \end{aligned}$$

What if this never ends? If it does, is that the only solution? Is it even a solution itself??

(N.B.: $x = x+1 \Rightarrow x^2 = x^2 + 2x + 1 \Rightarrow x = -\frac{1}{2}$

But this is not a solution!)

We want a strategy that is always <u>completely conclusive</u>.

Idea: Think of the system as being a single thing, instead of being made up of individual equations.

Def: Two systems are <u>equivalent</u> if they have identical solution sets.

Def: The following are called <u>elementary equation operations</u>:

1. interchanging two equations
2. adding a multiple of one equation to another
3. multiplying an equation by a nonzero number

These are not just algebraically legal — they are also <u>reversible</u>.

So,

Thm: If S_1 is acted on by elementary operations to yield a new system S_2, then S_1 and S_2 are equivalent.

Ex:

$$S_1 \begin{cases} x_1 + 3x_2 = 5 \\ 2x_1 - x_2 = 3 \end{cases}$$

$$\begin{cases} x_1 + 3x_2 = 5 & ① \\ -7x_2 = -7 & ② - 2① \end{cases}$$

$$\begin{cases} x_1 + 3x_2 = 5 & ① \\ x_2 = 1 & -\tfrac{1}{7}② \end{cases}$$

$$S_2 \begin{cases} x_1 = 2 & ① - 3② \\ x_2 = 1 & ② \end{cases}$$

By our theorem, sol. sets for S_1, S_2 are the same; but S_2's sol. set is obvious. ☺

<u>Notation</u>:
- "ⓝ" is a reference to an equation in the <u>previous</u> system, not the original
- Put formula for an equation next to that equation itself (book does differently...)

Matrix Notation

Note, if you fix the order of the variables in the equations, only the coefficients and constants change. This allows for a convenient short hand.

$$\left\{\begin{array}{c} a_{11}x_1 + \cdots + a_{1n}x_n = b_1 \\ \vdots \\ a_{m1}x_1 + \cdots + a_{mn}x_n = b_m \end{array}\right\} \text{ becomes } \left(\begin{array}{ccc|c} a_{11} & \cdots & a_{1n} & b_1 \\ \vdots & & \vdots & \vdots \\ a_{m1} & \cdots & a_{mn} & b_m \end{array}\right)$$

This is called an augmented matrix.

The vertical bar is just a reference point, here indicating the "=".

The left side is the coefficient matrix.

The right side is the vector of constants.

Using this notation, "equation operations" are "row operations".

Elementary Row Operations

1. interchanging two rows
2. adding a multiple of one row to another
3. multiplying a row by a nonzero number

Reduced Row Echelon Form

The process of solving a system comes down to applying row operations to make the matrix as convenient as possible... (?)

Def:) A matrix is in <u>row echelon form</u> if each row has more leading zeroes than the preceding row (or is all zeroes).

Def:) The first nonzero entry in a row is called a <u>pivot</u>.

Def:) A matrix (usually a coefficient matrix) is in <u>reduced row echelon form</u> if

1. it is in row echelon form
2. all pivots are 1
3. all other entries in a pivot column are zero.

(A system is in rref if its coefficient matrix is in rref, as above.)

Thm:) For any system of equations (and assuming an ordering of the variables):

1. the matrix <u>can be</u> reduced to rref
2. there is a <u>unique</u> rref
3. there is a "convenient" means of deducing the solutions from the rref.

Ex:) $\begin{pmatrix} 1 & 0 & 0 \\ 0 & 1 & 0 \\ 0 & 0 & 1 \end{pmatrix}$, $\begin{pmatrix} 1 & 3 & 0 \\ 0 & 0 & 1 \\ 0 & 0 & 0 \\ 0 & 0 & 0 \end{pmatrix}$, $\begin{pmatrix} 1 & 2 & 4 & 0 & 5 \\ 0 & 0 & 0 & 1 & 0 \end{pmatrix}$

Recall that, having fixed an order for the variables, columns correspond to variables.

Def:) A variable whose column (in rref) contains a pivot is a <u>pivot variable</u>. The other variables are called <u>free variables</u>.

Why is rref "convenient"? Because, when a matrix is in rref,

You can always solve for the pivot variables in terms of <u>only</u> the free variables

Ex:) $\left(\begin{array}{ccc|c} 1 & 3 & 0 & 2 \\ 0 & 0 & 1 & 5 \\ 0 & 0 & 0 & 0 \end{array}\right)$ Solving for the pivot variables (x_1, x_3) in terms of the free variable (x_2), we get:

$$\begin{aligned} x_1 &= 2-3x_2 \\ x_3 &= 5 \end{aligned} \implies \begin{pmatrix} x_1 \\ x_2 \\ x_3 \end{pmatrix} = \begin{pmatrix} 2-3x_2 \\ x_2 \\ 5 \end{pmatrix}$$

This always gives <u>exactly</u> the solution set.

Why does this give us the exact solution set?

1. Every thing you get is a solution, for every value of the free variables, because all of the "pivot equations" are satisfied — because we solved for the pivot variables in those equations in terms of the free variables.

2. Every solution will be found in this way, because

for any solution, the values of the free variables can be used to determine what the pivot variables in that solution would have to be — which we already found when we solved for the pivot variables in terms of the free variables

But watch out for contradictions!

Ex:

$$\left(\begin{array}{ccc|c} 1 & 3 & 0 & 2 \\ 0 & 0 & 1 & 5 \\ 0 & 0 & 0 & 7 \end{array}\right)$$

This last row represents the equation " $0 = 7$ "...

Of course there are no values of x_1, x_2, x_3 that make this true.

So the original system has no solutions

Gauss–Jordan Elimination

Usually the easiest way to get to rref is to "fix" the <u>columns</u> one at a time, from left to right.

Ex:
$$\left(\begin{array}{ccc|c} 1 & -2 & -1 & 1 \\ -3 & 8 & 7 & -5 \\ -4 & 11 & 11 & -8 \end{array}\right)$$

The problem with the first column is that the pivot (the 1 in the first row) is not the only nonzero entry.

$$\left(\begin{array}{ccc|c} 1 & -2 & -1 & 1 \\ 0 & 2 & 4 & -2 \\ 0 & 3 & 7 & -4 \end{array}\right) \begin{array}{l} ① \\ ② + 3① \\ ③ + 4① \end{array}$$

The next pivot will be in the second column, second row; we have to make it a "1" and eliminate the other entries.

$$\left(\begin{array}{ccc|c} 1 & -2 & -1 & 1 \\ 0 & 1 & 2 & -1 \\ 0 & 3 & 7 & -4 \end{array}\right) \begin{array}{l} ① \\ ②/2 \\ ③ \end{array}$$

$$\left(\begin{array}{ccc|c} 1 & 0 & 3 & -1 \\ 0 & 1 & 2 & -1 \\ 0 & 0 & 1 & -1 \end{array}\right) \begin{array}{l} ① + 2② \\ ② \\ ③ - 3② \end{array}$$

Finally, we fix the last column.

$$\left(\begin{array}{ccc|c} 1 & 0 & 0 & 2 \\ 0 & 1 & 0 & 1 \\ 0 & 0 & 1 & -1 \end{array}\right) \begin{array}{l} ① - 3③ \\ ② - 2③ \\ ③ \end{array}$$

Notes

① Sometimes there is no pivot that can be arranged in a given column... No problem!

Ex: $\begin{pmatrix} 1 & 0 & 2 & 7 \\ 0 & 1 & 4 & 8 \\ 0 & 0 & 0 & 5 \\ 0 & 0 & 0 & 2 \end{pmatrix}$ can't get a pivot into this column.

② Sometimes you have to, or simply might want to switch rows to get a pivot where it needs to be

Ex: $\begin{pmatrix} 0 & 1 & 2 \\ 5 & 2 & 4 \\ 3 & 0 & 6 \end{pmatrix}$ Could switch ①, ②... but better to switch ①, ③ since it will avoid, or at least delay, the appearance of fractions.

Ex: $\begin{pmatrix} 5 & 1 & 8 \\ 2 & 3 & 1 \\ 1 & 2 & 3 \end{pmatrix}$ Switching ①, ③ not necessary, but gets a 1 to the top left position without fractions.

③ Can do other things to avoid fractions...

$$\begin{pmatrix} 4 & 2 & 7 \\ 8 & 5 & 1 \\ 3 & 0 & 4 \end{pmatrix}$$

$$\begin{pmatrix} 1 & 2 & 3 \\ 8 & 5 & 1 \\ 3 & 0 & 4 \end{pmatrix} \begin{matrix} ① - ③ \\ ② \\ ③ \end{matrix}$$

Combining Row Operations

$$\begin{pmatrix} 1 & 1 \\ 1 & 2 \end{pmatrix}$$

$$\begin{pmatrix} 1 & 1 \\ 0 & 1 \end{pmatrix} \begin{matrix} \textcircled{1} \\ \textcircled{2} - \textcircled{1} \end{matrix}$$

$$\begin{pmatrix} 1 & 0 \\ 0 & 1 \end{pmatrix} \begin{matrix} \textcircled{1} - \textcircled{2} \\ \textcircled{2} \end{matrix}$$

It is tempting to combine these two steps into a single step:

$$\Big) \begin{matrix} \textcircled{1} - \textcircled{2} \\ \textcircled{2} - \textcircled{1} \end{matrix}$$

But that is not allowed in this case!

In fact in this case doing so would lead to a non-equivalent system!

$$\begin{pmatrix} 1 & 1 \\ 1 & 2 \end{pmatrix}$$

$$\begin{pmatrix} 0 & -1 \\ 0 & 1 \end{pmatrix} \begin{matrix} \textcircled{1} - \textcircled{2} \\ \textcircled{2} - \textcircled{1} \end{matrix}$$

← this further reduces to the non-equivalent: $\begin{pmatrix} 0 & 1 \\ 0 & 0 \end{pmatrix}$

What went wrong? The point here is that in the original reduction, the "$\textcircled{1} - \textcircled{2}$" refers to a row $\textcircled{2}$ that is <u>not the same</u> as the original row $\textcircled{2}$.

Combining these row operations into a single step blurs this distinction; so, it is <u>not</u> allowed.

As stated, row operations should be done one-at-a-time.

But — there are cases where combining row operations does not lead to any problems. Most importantly:

> If you leave one row <u>fixed</u>, and use only that row to adjust other rows, no problems will arise.

Ex:

$$\begin{pmatrix} 1 & 2 & 0 \\ 2 & 3 & 1 \\ 4 & 9 & 6 \end{pmatrix}$$

$$\begin{pmatrix} 1 & 2 & 0 \\ 0 & -1 & 1 \\ 0 & 1 & 6 \end{pmatrix} \begin{matrix} \textcircled{1} \\ \textcircled{2} - 2\textcircled{1} \\ \textcircled{3} - 4\textcircled{1} \end{matrix}$$

this is left fixed ($\textcircled{1}$)

only the fixed $\textcircled{1}$ is used to adjust the other rows.

Variation:

$$\begin{pmatrix} 2 & 1 & 4 \\ 3 & 2 & 16 \\ 5 & 8 & 1 \end{pmatrix}$$

$$\begin{pmatrix} 1 & 1/2 & 2 \\ 0 & 1 & 20 \\ 0 & 11 & -18 \end{pmatrix} \begin{matrix} \textcircled{1}/2 \\ 2\textcircled{2} - 3\textcircled{1} \\ 2\textcircled{3} - 5\textcircled{1} \end{matrix}$$

Here, the first row is only scaled. Had we done that scaling in an initial step, that would only change the multiples needed to use it to adjust the other rows.

Existence, Uniqueness, and Rank

Def:) The rank of a matrix is the number of pivots in the rref.

Rank relates to two natural questions about the solutions to a system of equations.

Existence

We have already seen that a system has solutions iff there are no contradictions.

$$\left(\begin{array}{ccccc|c} 1 & 2 & 0 & 5 & 7 & ? \\ 0 & 0 & 1 & 2 & 8 & ? \\ 0 & 0 & 0 & 0 & 0 & \boxed{?} \\ 0 & 0 & 0 & 0 & 0 & \boxed{?} \end{array}\right)$$

If these are both zero, solutions exist. If either is nonzero, no solutions.

rows of zeroes in the coefficient matrix

What can we deduce about this if we know only the coefficient matrix?

Def:) A coefficient matrix has the existence property (general existence, universal existence, ...) if solutions must exist no matter what the constants are on the right side.

How can you tell if a matrix has existence?

Thm:) A coefficient matrix has the existence property if

(1) there are <u>no</u> rows of zeroes in rref

(equiv:) (2) there is a pivot in every row of rref

(3) rank = # rows

(If there were a row of zeroes, there might be a contradiction!)

Ex:) Say $\text{rref}(A) = \begin{pmatrix} 1 & 3 & 0 \\ 0 & 0 & 1 \end{pmatrix}$, $\text{rref}(B) = \begin{pmatrix} 0 & 1 & 2 \\ 0 & 0 & 0 \\ 0 & 0 & 0 \end{pmatrix}$

A <u>does</u> have universal existence; B <u>does not</u>.

<u>Related point:</u>) A system is called <u>homogeneous</u> if the right side is all zeroes:

$$\begin{aligned} a_{11}x_1 + \cdots + a_{1n}x_n &= 0 \\ &\vdots \\ a_{m1}x_1 + \cdots + a_{mn}x_n &= 0 \end{aligned}$$

Note, a homogeneous system always has solutions, even though the coefficient matrix might not have universal existence.

The solution $x_1 = 0, \ldots, x_n = 0$ is called the <u>trivial solution</u>.

Uniqueness

(NB— the uniqueness question should be thought of as independent of the existence question. Ignore potential existence problems; that is, act as if a solution is known to exist, even if that is not known.)

What can we deduce about uniqueness of (possible) solutions if we know only the coefficient matrix?

Def:) A coefficient matrix has the uniqueness property if (possible) solutions must be unique.

How can you tell if a matrix has uniqueness?

Thm:) A coefficient matrix has the uniqueness property if

1. there are no free variables in rref
2. (equiv:) there is a pivot in every column of rref
3. rank A = # cols

(If there were free variables, we could not have unique solutions...)

Ex:) $\text{rref}(A) = \begin{pmatrix} 1 & 0 \\ 0 & 0 \\ 0 & 0 \end{pmatrix}$, $\text{rref}(B) = \begin{pmatrix} 1 & 0 \\ 0 & 1 \\ 0 & 0 \end{pmatrix}$

A does not have uniqueness; B does

Related Facts

Obs:) If A is $m \times n$, then $\text{rank}(A) \le m, n$.
(rows) ↗ ↖ (cols)

Thm:) If $m > n$ (more eqs than vars; a "tall & thin" matrix), then A cannot have universal existence.

(There might be RHS constants where solutions exist... but there will be RHS constants where solutions DNE.)

Thm:) If $m < n$ (more vars than eqs; a "short + wide" matrix), then A cannot have uniqueness.

Thm:) A linear system must have either 0, 1, or ∞'ly many solutions.

1.2 - Matrices and Matrix Operations

Previously we have seen matrices used as a shorthand for systems of equations. However, they can be used for other things — and can be thought of as independent things.

Matrix Addition

Say $A = (a_{ij})$, $B = (b_{ij})$ have same dimensions.

Then $C = A + B$ where $C = (c_{ij})$

is defined by $c_{ij} = a_{ij} + b_{ij}$

Ex: $\begin{pmatrix} 1 & 3 \\ 7 & 2 \end{pmatrix} + \begin{pmatrix} 2 & -1 \\ -3 & 5 \end{pmatrix} = \begin{pmatrix} 3 & 2 \\ 4 & 7 \end{pmatrix}$

Scalar Multiplication

Say $A = (a_{ij})$. Then

$C = kA$ where $C = (c_{ij})$

is defined by $c_{ij} = k\,a_{ij}$

Ex: $5\begin{pmatrix} 3 & 2 \\ 1 & -4 \end{pmatrix} = \begin{pmatrix} 15 & 10 \\ 5 & -20 \end{pmatrix}$

Matrix Multiplication

Need ① # cols of A = # rows of B

(equiv)

② # elts of row of A = # elts of col of B

$$A = \begin{pmatrix} \text{---} A_1 \text{---} \\ \vdots \\ \text{---} A_m \text{---} \end{pmatrix} = \begin{pmatrix} | & & | \\ a_1 & \cdots & a_n \\ | & & | \end{pmatrix} \quad \text{is } m \times n$$

$$B = \begin{pmatrix} \text{---} B_1 \text{---} \\ \vdots \\ \text{---} B_n \text{---} \end{pmatrix} = \begin{pmatrix} | & & | \\ b_1 & \cdots & b_p \\ | & & | \end{pmatrix} \quad \text{is } n \times p$$

Then $C = AB$, $C = (c_{ij})$ is defined by a dot product:

$$c_{ij} = A_i \cdot b_j$$

Note the dimensions:

$$\underset{(m \times p)}{C} = \underset{(m \times n)}{A} \; \underset{(n \times p)}{B}$$

Ex:

$$\begin{pmatrix} 1 & 3 & 1 \\ 2 & 0 & -1 \end{pmatrix} \begin{pmatrix} 2 & 3 \\ 5 & 1 \\ 4 & -2 \end{pmatrix} = \begin{pmatrix} 21 & 4 \\ 0 & 8 \end{pmatrix}$$

Alternatives:

① Cols of AB are linear combinations of cols of A, using cols of B as coeffs.

$$\begin{pmatrix} 1 & 2 \\ 3 & 4 \end{pmatrix}\begin{pmatrix} 5 & 6 \\ 7 & 8 \end{pmatrix} = \begin{pmatrix} 19 & 22 \\ 43 & 50 \end{pmatrix}$$

$$= 5\begin{pmatrix} 1 \\ 3 \end{pmatrix} + 7\begin{pmatrix} 2 \\ 4 \end{pmatrix} = \begin{pmatrix} 19 \\ 43 \end{pmatrix}$$

② Rows of AB are linear combinations of rows of B, using rows of A as coeffs.

$$\begin{pmatrix} 1 & 2 \\ 3 & 4 \end{pmatrix}\begin{pmatrix} 5 & 6 \\ 7 & 8 \end{pmatrix} = \begin{pmatrix} 19 & 22 \\ 43 & 50 \end{pmatrix}$$

$$3\begin{pmatrix} 5 & 6 \end{pmatrix} + 4\begin{pmatrix} 7 & 8 \end{pmatrix} = \begin{pmatrix} 43 & 50 \end{pmatrix}$$

③ Sigma notation:

$A = (a_{ij})$, $B = (b_{ij})$, $C = (c_{ij}) = AB$, then

$$c_{ij} = \sum_{k=1}^{n} a_{ik} b_{kj}$$

Properties

$A(BC) = (AB)C$ ← (this one is hard to prove here; but will be easier with "linear transformations".)

$A(B+C) = AB + AC$

$k(AB) = (kA)B = A(kB)$

} these you can prove now!

Defs: $O_{m\times n} = \begin{pmatrix} 0 & \cdots & 0 \\ \vdots & & \vdots \\ 0 & \cdots & 0 \end{pmatrix}$ ← $(m\times n)$

$I_n = \begin{pmatrix} 1 & & O \\ & \ddots & \\ O & & 1 \end{pmatrix}$ ← $(n\times n)$

Thm: $A + O = A = O + A$

Thm: If A is $m\times n$, then

$$I_m A = A = A I_n$$

Outline of Pf: Think of products as l.c.'s of rows, cols.

$$\begin{pmatrix} 1 & & & \\ & 1 & & \\ & & \ddots & \\ & & & 1 \end{pmatrix} \begin{pmatrix} \text{---} A_1 \text{---} \\ \text{---} A_2 \text{---} \\ \vdots \\ \text{---} A_m \text{---} \end{pmatrix} = \begin{pmatrix} & & \\ & & \\ & & \end{pmatrix}$$

Obs: You can view a system of equations as a single matrix equation.

Consider

$$\begin{aligned} a_{11}x_1 + \cdots + a_{1n}x_n &= b_1 \\ &\vdots \\ a_{m1}x_1 + \cdots + a_{mn}x_n &= b_m \end{aligned}$$

and choose

$$A = \begin{pmatrix} a_{11} & \cdots & a_{1n} \\ \vdots & & \vdots \\ a_{m1} & \cdots & a_{mn} \end{pmatrix}, \quad X = \begin{pmatrix} x_1 \\ \vdots \\ x_n \end{pmatrix}, \quad B = \begin{pmatrix} b_1 \\ \vdots \\ b_m \end{pmatrix}$$

Then

$$AX = \qquad \text{and} \qquad B =$$

So the entire system is represented by

$$AX = B$$

Thm:) Suppose that X_p is any solution to $AX=B$. ("particular solution")

Then X is a solution iff $X = X_p + X_H$, where

X_H is a solution to $AX=0$ ("homogeneous solution").

Pf:) First observe that if $X = X_p + X_H$, then

$$AX = AX_p + AX_H = B + 0 = B$$

So X is a solution.

On the other hand, if X is a solution, then

$$A(X - X_p) = AX - AX_p = B - B = 0$$

which means $X_H = X - X_p$ is a homogeneous solution

and thus $X = X_p + X_H$.

<u>Interpretation</u>: If you know <u>any</u> particular solution, and <u>all</u> homogeneous solutions, then the sum is the complete set of sols.

Ex:) Consider $\left(\begin{array}{ccc|c} 1 & 2 & -2 & -5 \\ -2 & -4 & 5 & 13 \\ -5 & -10 & 7 & 16 \end{array}\right)$

Can check $\begin{pmatrix} -2y \\ y \\ 0 \end{pmatrix}$ is all homogeneos sols, and $\begin{pmatrix} 1 \\ 0 \\ 3 \end{pmatrix}$ is a particular sol.

So the complete set of sols is

$$\begin{pmatrix} 1 \\ 0 \\ 3 \end{pmatrix} + \begin{pmatrix} -2y \\ y \\ 0 \end{pmatrix} = \begin{pmatrix} 1-2y \\ y \\ 3 \end{pmatrix}$$

Thm:) If M has the uniqueness property, then you can "left cancel" with M. That is, $(MS = MT) \Rightarrow (S = T)$

Pf:) For all $\vec{x}$, we have

$$MS\vec{x} = MT\vec{x}$$

$$M(S\vec{x}) = M(T\vec{x}) = \vec{v}$$

Let $S\vec{x} = \vec{y}_1$, $T\vec{x} = \vec{y}_2$, so the above becomes

$$M\vec{y}_1 = M\vec{y}_2 = \vec{v}$$

Then $\vec{y}_1, \vec{y}_2$ are both solutions to

$$M\vec{y} = \vec{v}$$

Since M has the uniqueness property, we get

$$\vec{y}_1 = \vec{y}_2$$

$$S\vec{x} = T\vec{x}$$

Since this is true for all $\vec{x}$, we have $S = T$.

Thm:) If N has the existence property, then you can "right cancel" with N.

That is, $(SN = TN) \Rightarrow (S = T)$

Pf:) We start with $SN\vec{x} = TN\vec{x}$

Let $N\vec{x} = \vec{b}$; the existence property of N means that $\vec{b}$ can be any vector.

Substituting above, we get that

$$S\vec{b} = T\vec{b}$$

for all $\vec{b}$, and so $S = T$.

Inverses of Matrices

For an $n \times n$ matrix A, we say B is an <u>inverse</u> if

$$AB = BA = I$$

Ex:) $A = \begin{pmatrix} 2 & 1 \\ 3 & 2 \end{pmatrix}$ $\quad B = \begin{pmatrix} 2 & -1 \\ -3 & 2 \end{pmatrix}$

Note $\begin{pmatrix} 2 & 1 \\ 3 & 2 \end{pmatrix}\begin{pmatrix} 2 & -1 \\ -3 & 2 \end{pmatrix} = \begin{pmatrix} 1 & 0 \\ 0 & 1 \end{pmatrix} = \begin{pmatrix} 2 & -1 \\ -3 & 2 \end{pmatrix}\begin{pmatrix} 2 & 1 \\ 3 & 2 \end{pmatrix}$

If A has an inverse, say A is <u>invertible</u>.

Thm:) If A is invertible, then the inverse is <u>unique</u>.

Pf:) Say B, C are both inverses for A. Then

$$BAC = (BA)C = IC = C$$
$$BAC = B(AC) = BI = B$$

So $B = C$. (Given this uniqueness, we write the inverse as A^{-1}.)

Thm:) If A is $n \times n$ and $AB = I$, then $BA = I$ (so A is invertible). (Note, A must be square!)

Pf:) Say $AB = I$. First we will show A has the uniqueness property.

The system $A\vec{y} = \vec{c}$ always has solutions, because $A(B\vec{c}) = (AB)\vec{c} = I\vec{c} = \vec{c}$.

So A has the existence property

$\Rightarrow$ pivot in every row

$\Rightarrow$ pivot in every column (b/c A is square!)

$\Rightarrow$ A has the uniqueness property

Then $AB = I \Rightarrow (AB)A = (I)A$

$\Rightarrow A(BA) = A(I)$

and we can left cancel the A because of the uniqueness property. So $BA = I$

Elementary Matrices

An elementary matrix is the result of applying a row operation to the identity matrix.

Let's define a row operation matrix to be a matrix F that executes a row operation (by left multiplication):

$$A \overset{\text{row op}}{\rightsquigarrow} FA$$

Ex:

$$\begin{pmatrix} 1 & 0 & 0 \\ 5 & 1 & 0 \\ 0 & 0 & 1 \end{pmatrix}\begin{pmatrix} 1 & 0 & 2 & 1 \\ 3 & 4 & 3 & 7 \\ 2 & -3 & -2 & -1 \end{pmatrix} = \begin{pmatrix} 1 & 0 & 2 & 1 \\ 8 & 4 & 13 & 12 \\ 2 & -3 & -2 & -1 \end{pmatrix}\begin{matrix} (1) \\ (2)+5(1) \\ (3) \end{matrix}$$

Ex:

$$\begin{pmatrix} 0 & 0 & 1 \\ 0 & 1 & 0 \\ 1 & 0 & 0 \end{pmatrix}\begin{pmatrix} 1 & 2 \\ 3 & 4 \\ 5 & 6 \end{pmatrix} = \begin{pmatrix} 5 & 6 \\ 3 & 4 \\ 1 & 2 \end{pmatrix}\begin{matrix} (3) \\ (2) \\ (1) \end{matrix}$$

Note: (1) For <u>every</u> row operation, each new row is a linear combination of the previous rows (of A).

(2) When left multiplying a matrix A by a matrix F (as in FA), the rows of the product are linear combinations of the rows of A.

(3) This similarity shows these row operation matrices always exist.

So:

$$A \xrightarrow{\text{row op.}} B$$

$$FA = B$$

same!

Thm:) If E is obtained by a row operation on I, then the result of applying the same row operation on a matrix A is EA.

Pf:) Let F be the row operation matrix for this operation; then

$$E = FI$$

so

$$E = F$$

Applying this operation to A gives

$$FA = EA.$$

Note, a different way to view this is simply that elementary matrices and "row operation matrices" are the same thing. We will refer to these henceforth as elementary matrices.

Thm:) A is invertible $\iff$ $\operatorname{rref}(A) = I$ (A is "nonsingular")

Pf:) ($\Rightarrow$) Say A is invertible. We need to show there is a pivot in every row (since A is square, it will follow that there is a pivot in every column.)

That is, we need to show existence for $A\vec{x} = \vec{b}$.

This is easy, since $\vec{x} = A^{-1}\vec{b}$ is clearly a solution.

($\Leftarrow$) Say rref(A) is the identity. We can realize the row reduction as

$$E_k \cdots E_1 A = I$$

Writing $E = E_k \cdots E_1$, this becomes

$$EA = I$$

So $E = A^{-1}$ and A is invertible.

This also leads us to a method for finding A^{-1}...

$$E = A^{-1} \implies EI = A^{-1}$$

So if we apply E to the augmented matrix $(A \mid I)$, we get

$$E(A \mid I) = (EA \mid EI) = (I \mid A^{-1}).$$

That is — apply row ops to I <u>in parallel</u> as you do so for A; and if A reduces to I, then I reduces to A^{-1}.

Ex: Is $\begin{pmatrix} 3 & 1 \\ 5 & 2 \end{pmatrix}$ invertible, and if so, what is the inverse?

$$\left(\begin{array}{cc|cc} 3 & 1 & 1 & 0 \\ 5 & 2 & 0 & 1 \end{array}\right)$$

$$\left(\begin{array}{cc|cc} 3 & 1 & 1 & 0 \\ -1 & 0 & -2 & 1 \end{array}\right) \begin{matrix} ① \\ ② - 2① \end{matrix}$$

$$\left(\begin{array}{cc|cc} 1 & 0 & 2 & -1 \\ 3 & 1 & 1 & 0 \end{array}\right) \begin{matrix} -② \\ ① \end{matrix}$$

$$\left(\begin{array}{cc|cc} 1 & 0 & 2 & -1 \\ 0 & 1 & -5 & 3 \end{array}\right) \begin{matrix} ① \\ ② - 3① \end{matrix}$$

This is the inverse (must have I on left!)

This tells us $\begin{pmatrix} 3 & 1 \\ 5 & 2 \end{pmatrix}$ is invertible.

Other facts: ① $(AB)^{-1} = B^{-1}A^{-1}$

② If A not square, $AB = I \not\Rightarrow BA = I$

Ex: $\begin{pmatrix} 1 & 0 & 0 \\ 0 & 1 & 0 \end{pmatrix}\begin{pmatrix} 1 & 0 \\ 0 & 1 \\ 0 & 0 \end{pmatrix} = \begin{pmatrix} 1 & 0 \\ 0 & 1 \end{pmatrix} = I_2$

$\begin{pmatrix} 1 & 0 \\ 0 & 1 \\ 0 & 0 \end{pmatrix}\begin{pmatrix} 1 & 0 & 0 \\ 0 & 1 & 0 \end{pmatrix} = \begin{pmatrix} 1 & 0 & 0 \\ 0 & 1 & 0 \\ 0 & 0 & 0 \end{pmatrix} \neq I_3$

Another application of elementary matrices involves solving the system $A\vec{x} = \vec{b}$ with several different $\vec{b}$ vectors.

Note, $A\vec{x} = \vec{b}_1$ and $A\vec{x} = \vec{b}_2$ require doing the <u>same</u> row operations to solve (or to determine if there is no solution). Why do all this work multiple times?

Suppose the row reduction of A is represented by

$$EA = R$$

Then solving $\quad A\vec{x} = \vec{b} \quad$ involves row reducing, with E...

$$EA\vec{x} = E\vec{b}$$

$$R\vec{x} = E\vec{b}$$

This is equivalent, and already row reduced.

Ex i) Solve $A\vec{x} = \vec{b}$ with $\vec{b} = \vec{b}_1, \vec{b}_2$, and $\vec{b}_3$.

① Reduce $(A \mid I)$ to $(R \mid E)$

② Solve $A\vec{x} = \vec{b}_1$ with $R\vec{x} = E\vec{b}_1$

Solve $A\vec{x} = \vec{b}_2$ with $R\vec{x} = E\vec{b}_2$

Solve $A\vec{x} = \vec{b}_3$ with $R\vec{x} = E\vec{b}_3$

⋮

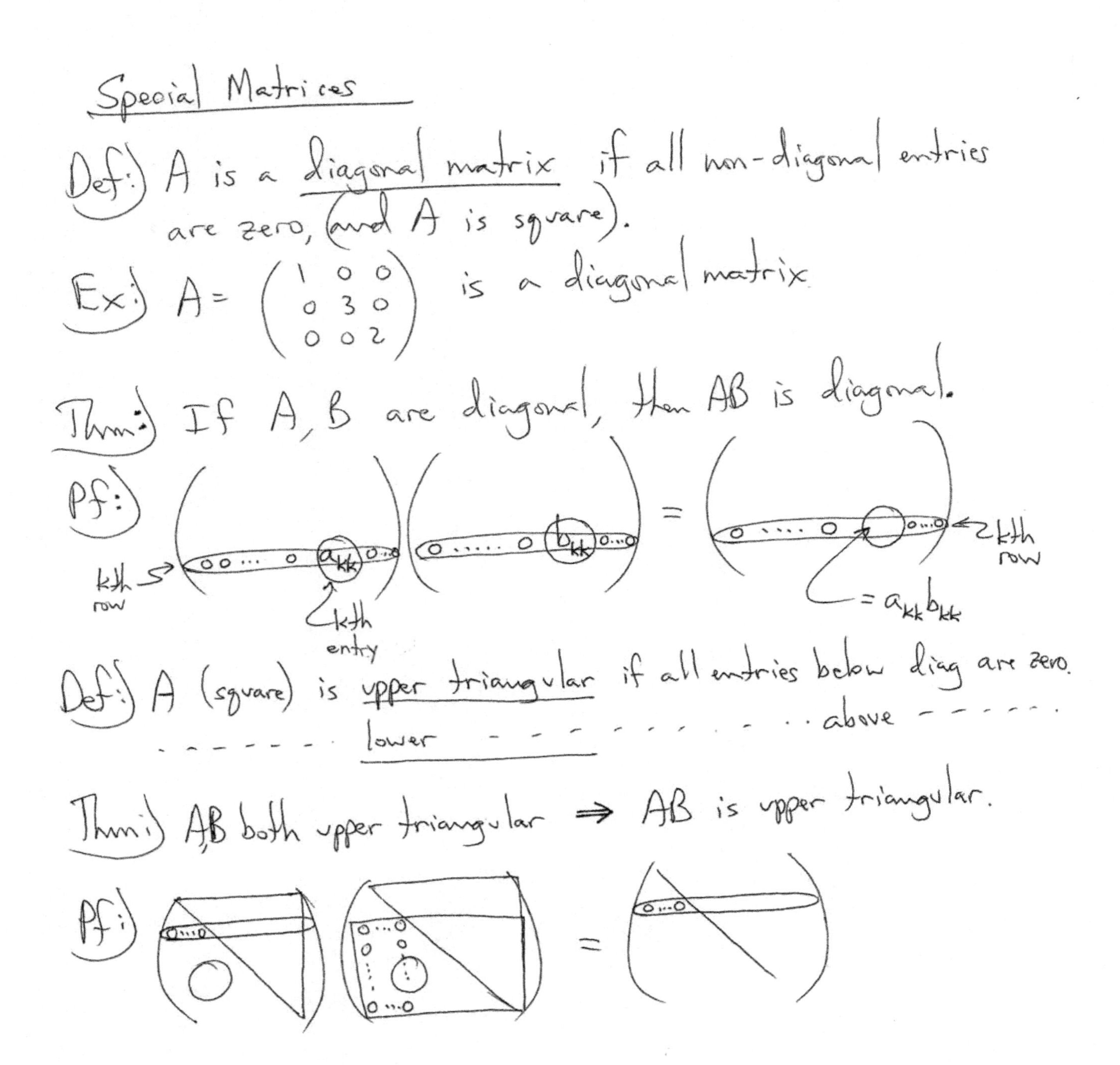
Special Matrices
Def:) A is a diagonal matrix if all non-diagonal entries are zero, (and A is square).
Ex:) A = (1 0 0 / 0 3 0 / 0 0 2) is a diagonal matrix
Thm:) If A, B are diagonal, then AB is diagonal.
Pf:)
kth row
kth entry
= akk bkk
Def:) A (square) is upper triangular if all entries below diag are zero.
lower . . . above
Thm:) A,B both upper triangular ⇒ AB is upper triangular.
Pf:)

Def: The <u>transpose</u>, A^T, of a matrix A is defined by

$$(A^T)_{ij} = A_{ji}$$

Ex: $A = \begin{pmatrix} 1 & 2 & 3 \\ 4 & 5 & 6 \end{pmatrix} \qquad A^T = \begin{pmatrix} 1 & 4 \\ 2 & 5 \\ 3 & 6 \end{pmatrix}$

Thm: $(AB)^T = B^T A^T$

Pf: $((AB)^T)_{ij} = (AB)_{ji} = A_j \cdot b_i$

$(B^T A^T)_{ij} = (i\text{th row of } B^T) \cdot (j\text{th col. of } A^T)$

$= (i\text{th col. of } B) \cdot (j\text{th row of } A)$

$= b_i \cdot A_j$ ✓

Thm: $(A^T)^{-1} = (A^{-1})^T$

Pf: $(A^{-1})^T A^T = (AA^{-1})^T = I^T = I$

Def: A (square) is <u>symmetric</u> if

$$A^T = A$$

Ex: $\begin{pmatrix} 1 & 3 & 7 \\ 3 & 10 & 2 \\ 7 & 2 & 4 \end{pmatrix}$ is symmetric $(a_{ij} = a_{ji}$ for all $i,j)$

Thm: A^TA, AA^T are symmetric

Pf: $(A^TA)^T = (A)^T(A^T)^T = A^TA$

$(AA^T)^T = (A^T)^T(A)^T = AA^T$

Thm: If A is invertible and symmetric, then A^{-1} is symm.

Pf: $(A^{-1})^T = (A^T)^{-1} = A^{-1}$

Determinants

Def:) The <u>ij-minor</u> of the $n \times n$ matrix A is the $(n-1)\times(n-1)$ matrix M_{ij} obtained by crossing out the ith row and jth column of A.

Ex:) $A = \begin{pmatrix} 1 & 2 & 3 \\ 4 & 5 & 6 \\ 7 & 8 & 9 \end{pmatrix}$ $M_{23} = \begin{pmatrix} 1 & 2 & \not{3} \\ \not{4} & \not{5} & \not{6} \\ 7 & 8 & \not{9} \end{pmatrix} = \begin{pmatrix} 1 & 2 \\ 7 & 8 \end{pmatrix}$

Def:) The <u>determinant</u> of a matrix is defined recursively:

- for a 1×1 matrix $A = (a)$, $\det(A) = a$
- for an $n \times n$ matrix $A = a_{ij}$,

$$\det A = \sum_{j=1}^{n} a_{ij} (-1)^{i+j} \det(M_{ij})$$

Ex:) $\det\begin{pmatrix} a & b \\ c & d \end{pmatrix} = a(-1)^{1+1}\det(d) + b(-1)^{1+2}\det(c)$
$= ad - bc$

Ex: $\det \begin{pmatrix} 1 & 2 & 3 \\ 4 & 5 & 6 \\ 7 & 8 & 9 \end{pmatrix}$

$$= 1(-1)^{1+1} \det \begin{pmatrix} 5 & 6 \\ 8 & 9 \end{pmatrix} + 2(-1)^{1+2} \det \begin{pmatrix} 4 & 6 \\ 7 & 9 \end{pmatrix} + 3(-1)^{1+3} \det \begin{pmatrix} 4 & 5 \\ 7 & 8 \end{pmatrix}$$

$$= 0$$

Thm: $\det A = \sum_{j=1}^{n} a_{ij} (-1)^{i+j} \det(M_{ij})$ (for any i)

$$= \sum_{i=1}^{n} a_{ij} (-1)^{i+j} \det(M_{ij}) \quad \text{(for any } j\text{)}$$

So you can compute determinants "along any row or column".

Note that $(-1)^{i+j}$ creates a checkerboard pattern on the matrix:

$(-1)^{1+1} = +1$ $\qquad$ $(-1)^{2+3} = -1$

$$\begin{matrix} \textcircled{+} & - & + & - & \\ - & + & \textcircled{-} & + & \cdots \\ + & - & + & - & \\ & & \vdots & & \end{matrix}$$

Ex:) $A = \begin{pmatrix} 1 & 2 & 3 \\ 4 & 5 & 6 \\ 7 & 8 & 9 \end{pmatrix}$

Along 2nd row:

$$\det A = (-1)\, 4 \det\begin{pmatrix} 2 & 3 \\ 8 & 9 \end{pmatrix} + (+1)\, 5 \det\begin{pmatrix} 1 & 3 \\ 7 & 9 \end{pmatrix} + (-1)\, 6 \det\begin{pmatrix} 1 & 2 \\ 7 & 8 \end{pmatrix}$$

Along 3rd column:

$$\det A = (+1)\, 3 \det\begin{pmatrix} 4 & 5 \\ 7 & 8 \end{pmatrix} + (-1)\, 6 \det\begin{pmatrix} 1 & 2 \\ 7 & 8 \end{pmatrix} + (+1)\, 9 \det\begin{pmatrix} 1 & 2 \\ 4 & 5 \end{pmatrix}$$

Def:) The ij-cofactor of A is

$$C_{ij} = (-1)^{i+j} \det M_{ij}$$

Then we can rewrite the previous formulas as

$$\det A = \sum_{j=1}^{n} a_{ij} C_{ij} \qquad (\text{for any } i)$$

$$= \sum_{i=1}^{n} a_{ij} C_{ij} \qquad (\text{for any } j)$$

Thm: $\det(A^T) = \det(A)$

Pf: We can show this by induction.

① Clearly this is true for 1×1 matrices

② If true for up to $(n-1)\times(n-1)$ matrices, then a cofactor expansion of $\det(A^T)$ along the kth row is

$$\det(A^T) = \sum_{l=1}^{n} a^T_{kl} (-1)^{k+l} \det(M'_{kl})$$

(M'_{kl}: kl-minor of A^T) (a^T_{kl}: kl entry of A^T)

$$= \sum_{l=1}^{n} a_{lk} (-1)^{l+k} \det(M_{lk}^T)$$

$$= \sum_{l=1}^{n} a_{lk} (-1)^{l+k} \det(M_{lk})$$

(because M_{lk} is $(n-1)\times(n-1)$)

$$= \det(A)$$

(this is the kth column cofactor expansion of $\det A$.)

Recall the following facts about determinants from multivariable calculus:

Thm:) Given $\vec{a}, \vec{b} \in \mathbb{R}^2$, form the matrix A and parallelogram P by

$$A = \begin{pmatrix} | & | \\ \vec{a} & \vec{b} \\ | & | \end{pmatrix}$$

Then ① $|\det A| = \text{area}(P)$

② $\text{sgn}(\det A)$ indicates order:

$\det A > 0 \iff \vec{a}, \vec{b}$ is a ccwise order

$\det A < 0 \iff \vec{a}, \vec{b}$ is a cwise order

$\det A = 0 \iff \vec{a}, \vec{b}$ in neither order

Thm:) Given $\vec{a}, \vec{b}, \vec{c} \in \mathbb{R}^3$, form the matrix A and parallelepiped R by

$$A = \begin{pmatrix} | & | & | \\ \vec{a} & \vec{b} & \vec{c} \\ | & | & | \end{pmatrix}$$

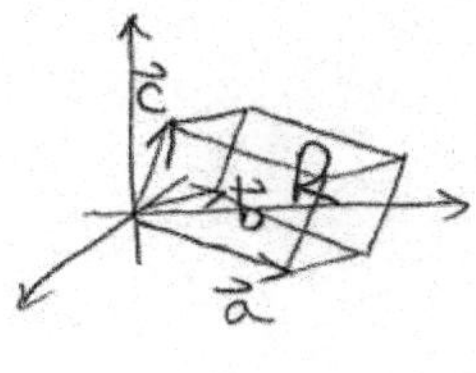

Then ① $|\det A| = \text{volume}(R)$

② $\text{sgn}(\det A)$ indicates order

$\det A > 0 \iff \vec{a}, \vec{b}, \vec{c}$ is a RH order

$\det A < 0 \iff \vec{a}, \vec{b}, \vec{c}$ is a LH order

$\det A = 0 \iff \vec{a}, \vec{b}, \vec{c}$ in neither order

Matrix multiplication allows us to reinterpret these results by thinking of a matrix as a function

Given $n \times n$ matrix A, let $T(\vec{x}) = A\vec{x}$

Then $T: \mathbb{R}^n \to \mathbb{R}^n$.

Note that

$$T(\vec{e}_i) = A\vec{e}_i = A\begin{pmatrix} 0 \\ \vdots \\ 0 \\ 1 \\ 0 \\ \vdots \\ 0 \end{pmatrix} = i\text{th column of } A$$

(1 in the ith position)

So, the <u>columns of A</u> are <u>the images of the standard basis vectors.</u>

We can use this result to reinterpret the parallelogram and parallelepiped as the images of the unit square and unit cube.

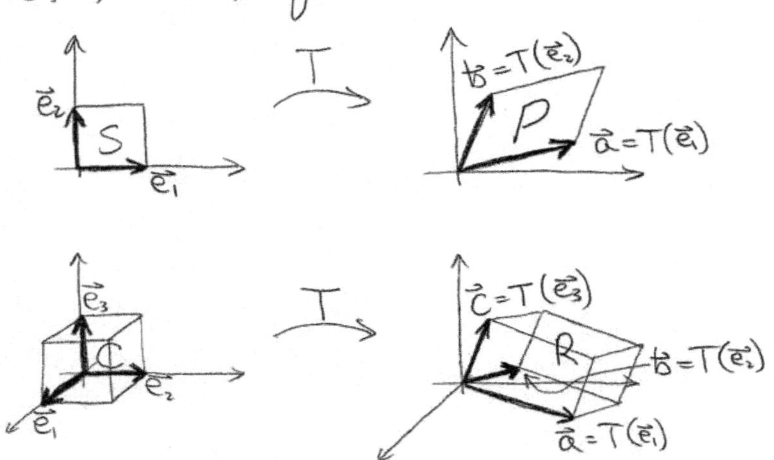

Then, for the 2×2 case,

$$|\det A| = \text{area}(P) = \frac{\text{area}(P)}{\text{area}(S)}$$

$= \text{factor by which } T \text{ stretches areas}$

(Relating to multivariable calculus, $J_T = A$.)

And since $\vec{e}_1, \vec{e}_2$ is a ccwise order, and $\text{sgn}(\det A)$ indicates whether $T(\vec{e}_1), T(\vec{e}_2)$ is ccwise or cwise, we can interpret that $\text{sgn}(\det A)$ indicates whether T "flips" the pair of vectors.

For the 3×3 case

$$|\det A| = \text{volume}(R) = \frac{\text{volume}(R)}{\text{volume}(C)}$$

$= \text{factor by which } T \text{ stretches volumes}$

(Again, $J_T = A$)

And $\text{sgn}(\det A)$ indicates whether T "reflects" the list of vectors.

More properties:

Thm: The determinant of a triangular matrix is the product of the diagonal entries.

Pf: Expand by cofactors along 1st row (for lower tri.) or 1st col. (for upper tri.) successively.

Thm: The determinant of a diagonal matrix is the product of the diagonal entries.

Pf: Follows from previous thm.

Geometric interpretation for 3×3 case with positive diagonal entries:

$$\det\begin{pmatrix} a & 0 & 0 \\ 0 & b & 0 \\ 0 & 0 & c \end{pmatrix} = \text{volume of}$$

Thm: If A has a row or column of zeroes, then $\det A = 0$

Pf: Cofactor expansion along that row or column.

Geometric interpretation for 3×3 case:

$$\det\begin{pmatrix} | & | & | \\ \vec{a} & \vec{b} & \vec{0} \\ | & | & | \end{pmatrix} = \text{volume of}$$

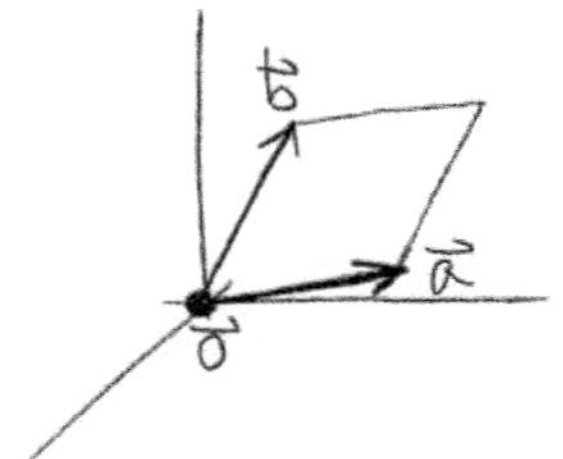

Multilinearity

A function $T : \mathbb{R}^n \to \mathbb{R}^m$ is <u>linear</u> iff

$$T(a\vec{x} + b\vec{y}) = aT(\vec{x}) + bT(\vec{y}) \quad \text{(for all } a, b, \vec{x}, \vec{y}\text{)}$$

Ex: A matrix function $T(\vec{x}) = A\vec{x}$ is linear

because

$$\begin{aligned} T(a\vec{x} + b\vec{y}) &= A(a\vec{x} + b\vec{y}) \\ &= A(a\vec{x}) + A(b\vec{y}) \\ &= aAx + bA\vec{y} \\ &= aT(\vec{x}) + bT(\vec{y}) \end{aligned}$$

Ex: $A = \begin{pmatrix} 1 & 0 & 3 \\ 2 & 1 & 0 \\ 0 & 1 & 1 \end{pmatrix}$, consider $\begin{pmatrix} 3 \\ 7 \\ 3 \end{pmatrix} = 3\begin{pmatrix} 1 \\ 2 \\ 0 \end{pmatrix} + 1\begin{pmatrix} 0 \\ 1 \\ 3 \end{pmatrix}$

$$A\begin{pmatrix} 3 \\ 7 \\ 3 \end{pmatrix} = 3A\begin{pmatrix} 1 \\ 2 \\ 0 \end{pmatrix} + 1A\begin{pmatrix} 0 \\ 1 \\ 3 \end{pmatrix}$$

$$\| \qquad\qquad\qquad \|$$

$$\begin{pmatrix} 12 \\ 13 \\ 10 \end{pmatrix} = 3\begin{pmatrix} 1 \\ 4 \\ 2 \end{pmatrix} + 1\begin{pmatrix} 9 \\ 1 \\ 4 \end{pmatrix}$$

Some functions have <u>several vector inputs</u>. For example, the input to the "det" function is a matrix – but a matrix can be thought of as several vectors (row or column).

$$A = \begin{pmatrix} \text{---} \vec{v}_1 \text{---} \\ \vdots \\ \text{---} \vec{v}_n \text{---} \end{pmatrix}$$

det A can be thought of as $\det(\vec{v}_1, \ldots, \vec{v}_n)$

A function is <u>multilinear</u> if it is linear in <u>each individual vector input</u> (while all other vector inputs are held constant).

<u>Thm:</u> Determinant is multilinear by rows and columns.

$$\det \begin{pmatrix} \text{---} \vec{v}_1 \text{---} \\ \vdots \\ \text{---} a\vec{v}_k + b\vec{w}_k \text{---} \\ \vdots \\ \text{---} \vec{v}_n \text{---} \end{pmatrix} = a \det \begin{pmatrix} \text{---} \vec{v}_1 \text{---} \\ \vdots \\ \text{---} \vec{v}_k \text{---} \\ \vdots \\ \text{---} \vec{v}_n \text{---} \end{pmatrix} + b \det \begin{pmatrix} \text{---} \vec{v}_1 \text{---} \\ \vdots \\ \text{---} \vec{w}_k \text{---} \\ \vdots \\ \text{---} \vec{v}_n \text{---} \end{pmatrix}$$

This can be proved by cofactor expansion along kth row (see 1112 Spring Exam1 Solutions.pdf, #5); and similarly for columns.

Ex:) Note that

$$\begin{pmatrix} -6 \\ 0 \\ 6 \end{pmatrix} = 10 \begin{pmatrix} 1 \\ 2 \\ 3 \end{pmatrix} - 4 \begin{pmatrix} 4 \\ 5 \\ 6 \end{pmatrix}$$

We can see the multilinearity of det in the second row by the easily checked fact that

$$\det \begin{pmatrix} 2 & 5 & 8 \\ -6 & 0 & 6 \\ 4 & 9 & 7 \end{pmatrix} = 10 \det \begin{pmatrix} 2 & 5 & 8 \\ 1 & 2 & 3 \\ 4 & 9 & 7 \end{pmatrix} - 4 \det \begin{pmatrix} 2 & 5 & 8 \\ 4 & 5 & 6 \\ 4 & 9 & 7 \end{pmatrix}$$

Similarly, we can observe the multilinearity of det in the third column by checking that

$$\det \begin{pmatrix} 3 & 4 & -6 \\ 5 & 1 & 0 \\ 7 & 1 & 6 \end{pmatrix} = 10 \det \begin{pmatrix} 3 & 4 & 1 \\ 5 & 1 & 2 \\ 7 & 1 & 3 \end{pmatrix} - 4 \det \begin{pmatrix} 3 & 4 & 4 \\ 5 & 1 & 5 \\ 7 & 1 & 6 \end{pmatrix}$$

Antisymmetry

Thm: If you switch two columns (or two rows) of a matrix, the determinant changes by a factor of (-1).

Ex:

$$\det\begin{pmatrix}1&0&2\\3&1&0\\0&1&4\end{pmatrix} = 1\det\begin{pmatrix}1&0\\1&4\end{pmatrix} - 0\det\begin{pmatrix}3&0\\0&4\end{pmatrix} + 2\det\begin{pmatrix}3&1\\0&1\end{pmatrix} = 10$$

$$\det\begin{pmatrix}1&2&0\\3&0&1\\0&4&1\end{pmatrix} = 1\det\begin{pmatrix}0&1\\4&1\end{pmatrix} - 2\det\begin{pmatrix}3&1\\0&1\end{pmatrix} + 0\det\begin{pmatrix}3&0\\0&4\end{pmatrix} = -10$$

Geometric interpretation in 3×3 case:

Both matrices represent same parallelepiped, same vol., with vectors in different order.

$$\begin{pmatrix}|&|&|\\\vec{a}&\vec{b}&\vec{c}\\|&|&|\end{pmatrix} \rightsquigarrow$$

$$\begin{pmatrix}|&|&|\\\vec{a}&\vec{c}&\vec{b}\\|&|&|\end{pmatrix} \rightsquigarrow$$

We also get a nice corollary.

Cor:) If two rows (or columns) of A are identical, then $\det A = 0$.

Pf:) Let A' be the result of switching the two identical rows (or columns).

Then by antisymmetry,

$$\det A' = -\det A$$

But $A' = A$, which also gives us

$$\det A' = \det A$$

So we must have $\det A = 0$.

Geometric interpretation in 3×3 case:

Row Operations and Elementary Matrices

Note, we already know how determinant is affected by 2 of 3 kinds of row operations. And multilinearity gives us

$$\det\begin{pmatrix} \vec{r}_1 \\ \vdots \\ \vec{r}_j + c\vec{r}_i \\ \vdots \\ \vec{r}_n \end{pmatrix} = 1 \det\begin{pmatrix} \vec{r}_1 \\ \vdots \\ \vec{r}_j \\ \vdots \\ \vec{r}_n \end{pmatrix} + c \det\begin{pmatrix} \vec{r}_1 \\ \vdots \\ \vec{r}_i \\ \vdots \\ \vec{r}_n \end{pmatrix}$$

jth row

(ith row, jth row are the same)

$$= 1 \det(A) \quad + c \cdot 0$$

Here then is a table of relationships between row ops and det:

Row op	Effect on det	Corresp. elem mat	Det of elem mat
Switch two rows	$\cdot(-1)$	$\begin{pmatrix} 1 & & & & \\ & \ddots & & & \\ & & 0 & 1 & \\ & & 1 & 0 & \\ & & & & \ddots \end{pmatrix}$	(-1)
Mult. row by $c \neq 0$	$\cdot\, c$	$\begin{pmatrix} 1 & & & & \\ & \ddots & & & \\ & & c & & \\ & & & \ddots & \\ & & & & 1 \end{pmatrix}$	c
Add mult. of one row to another	$\cdot\, 1$	$\begin{pmatrix} 1 & & c \\ & \ddots & \\ & & 1 \end{pmatrix}$	1

We can then use row reductions to compute determinants.

Ex:) $\begin{pmatrix} 0 & 1 & 3 \\ 1 & 0 & 2 \\ 0 & 2 & 10 \end{pmatrix} = A$

$\begin{pmatrix} 1 & 0 & 2 \\ 0 & 1 & 3 \\ 0 & 2 & 10 \end{pmatrix} \begin{matrix} ② \\ ① \\ ③ \end{matrix} \quad \leftarrow \quad \det = -\det A$

$\begin{pmatrix} 1 & 0 & 2 \\ 0 & 1 & 3 \\ 0 & 0 & 4 \end{pmatrix} \begin{matrix} ① \\ ② \\ ③ - 2② \end{matrix} \quad \leftarrow \quad \det = -\det A$

$\begin{pmatrix} 1 & 0 & 2 \\ 0 & 1 & 3 \\ 0 & 0 & 1 \end{pmatrix} \begin{matrix} ① \\ ② \\ ③/4 \end{matrix} \quad \leftarrow \quad \det = -\frac{1}{4} \det A$

$\begin{pmatrix} 1 & 0 & 0 \\ 0 & 1 & 0 \\ 0 & 0 & 1 \end{pmatrix} \begin{matrix} ① - 2③ \\ ② - 3② \\ ③ \end{matrix} \quad \leftarrow \quad \det = -\frac{1}{4} \det A$

$\det = 1$

So we solve to get $\det A = -4$.

This is <u>much more efficient</u> than cofactors for large matrices!

Thm:) For any elementary matrix E and any matrix A,

$$\det(EA) = \det(E)\det(A)$$

Pf:) Observation from previous table!

Cor:) The above is also true when E is a product of elementary matrices

Pf:) Successive applications of above thm. (Check!)

This allows us to relate determinants to nonsingularity / invertibility.

Thm:) A is nonsingular (invertible) $\iff$ $\det A \neq 0$

Pf:) Represent the row reduction of A by

$$EA = R$$

and thus

$$\det(E)\det(A) = \det(R)$$

From the table we know $\det(E)$ is never 0.

And $\det(R) \neq 0$ iff $R = I$.

Combining these observations gives the result.

This allows us to prove a very surprising result about determinants of products.

Thm: $\det(AB) = \det(A)\det(B)$

Pf: If A is invertible then $EA = I$, so $A = E^{-1}$ is a product of elementary matrices — for which we already know the result.

If A is not invertible, we know $\det(A) = 0$, so we need only show $\det(AB) = 0$ ...

In this case we have $EA = R$ where R has a row of zeroes. So $RB = EAB$ has a row of zeroes and thus determinant zero. But

$$0 = \det(EAB) = \det(E)\det(AB)$$

Since $\det(E)$ is not zero, we must have $\det(AB) = 0$.

We then have the following result about inverse matrices.

Thm: $\det(A^{-1}) = \dfrac{1}{\det(A)}$

Pf: $\det(A^{-1})\det(A) = \det(A^{-1}A) = \det(I) = 1$

Geometrically, we can see these results by interpreting matrices as functions $T : \mathbb{R}^n \to \mathbb{R}^n$ as previously discussed.

Thm:) A invertible $\iff$ $\det A \neq 0$

$\det \neq 0$
$\iff$ ∥piped "full"
$\iff$ T 1-1/onto

$\det = 0$
$\iff$ ∥piped "flat"
$\iff$ T not 1-1/onto

Thm:) $\det(AB) = \det(A)\det(B)$

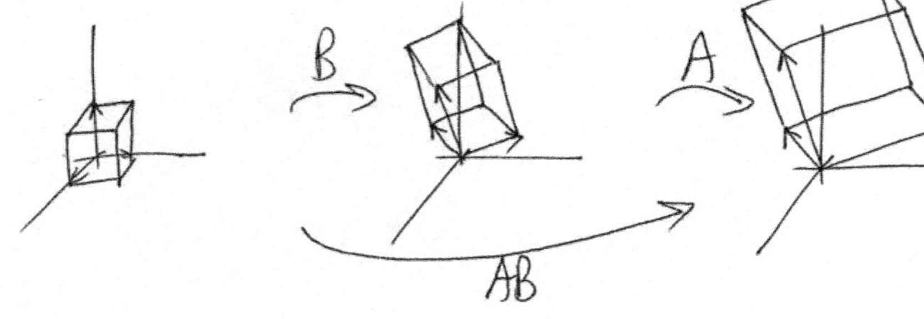

stretching factors multiply

Thm:) $\det(A^{-1}) = 1/\det(A)$

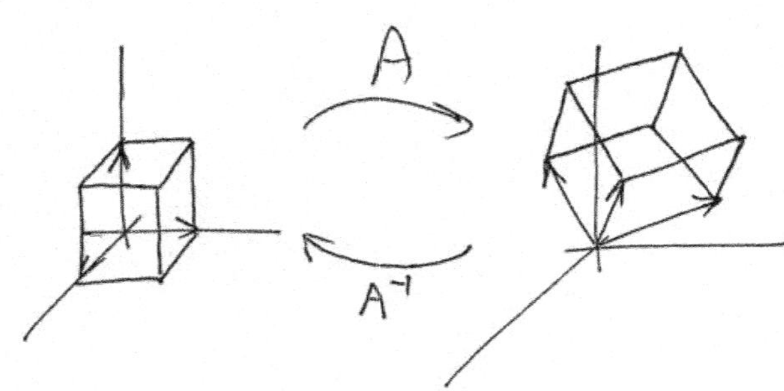

The inverse has to "unstretch" by however much A stretches.

More Applications of Determinants

Thm: Consider $A\vec{x} = \vec{b}$, with A square & invertible.

$$A = \begin{pmatrix} | & & | \\ \vec{a}_1 & \cdots & \vec{a}_n \\ | & & | \end{pmatrix}, \quad \vec{x} = \begin{pmatrix} x_1 \\ \vdots \\ x_n \end{pmatrix}$$

Define A_i to be result of replacing $\vec{a}_i$ with $\vec{b}$:

$$A_i = \begin{pmatrix} | & & | & & | \\ \vec{a}_1 & \cdots & \vec{b} & \cdots & \vec{a}_n \\ | & & | & & | \end{pmatrix}$$

Then $$x_i = \frac{\det A_i}{\det A} \qquad \text{(Cramer's rule)}$$

Pf: Interpret $A\vec{x} = \vec{b}$ as a statement about linear combinations of columns of A:

$$\vec{b} = x_1\vec{a}_1 + \cdots + x_n\vec{a}_n$$

Then

$$\det A_i = \det \begin{pmatrix} | & & & & | \\ \vec{a}_1 & \cdots & \underbrace{\left(x_1\vec{a}_1 + \cdots + x_n\vec{a}_n \right)}_{i\text{th column}} & \cdots & \vec{a}_n \\ | & & & & | \end{pmatrix}$$

$$= x_1 \det\begin{pmatrix} | & & | & & | \\ \vec{a}_1 & \cdots & \vec{a}_1 & \cdots & \vec{a}_n \\ | & & | & & | \end{pmatrix} + \cdots + x_i \det\begin{pmatrix} | & & | & & | \\ \vec{a}_1 & \cdots & \vec{a}_i & \cdots & \vec{a}_n \\ | & & | & & | \end{pmatrix} + \cdots + x_n \det\begin{pmatrix} | & & | & & | \\ \vec{a}_1 & \cdots & \vec{a}_n & \cdots & \vec{a}_n \\ | & & | & & | \end{pmatrix}$$

(first term: has 2 identical cols! ; middle term: $= A$; last term: has 2 identical cols!)

So $\quad \det A_i = x_i \det A$

Note, to solve an actual system, this is not very efficient.

But it can be useful if you need an equation for a single variable in the system.

Def: The <u>cofactor matrix</u> of A is the matrix C whose entries are the corresponding cofactors of A.

$$C = (c_{ij}), \quad c_{ij} = (-1)^{i+j} \det M_{ij}$$

Recall that the cofactor expansion of determinant is

$$\det A = \sum_{j=1}^{n} a_{ij} C_{ij} \quad (= (i\text{th row of } A) \cdot (i\text{th row of } C))$$

Def: The <u>adjoint</u> of A is the transpose of the cofactor matrix.

$$\operatorname{adj}(A) = C^T$$

Thm: $A^{-1} = \dfrac{\operatorname{adj}(A)}{\det(A)}$ $\quad$ (Equiv.: $A \operatorname{adj}(A) = \det(A) I$)

Pf: We compute $A \operatorname{adj}(A)$; the ij element is

$$\sum_{k=1}^{n} a_{ik} (\operatorname{adj}(A))_{kj}$$

$$= \sum_{k} a_{ik} C_{jk}$$

We consider separately the cases $i=j$ and $i \neq j$.

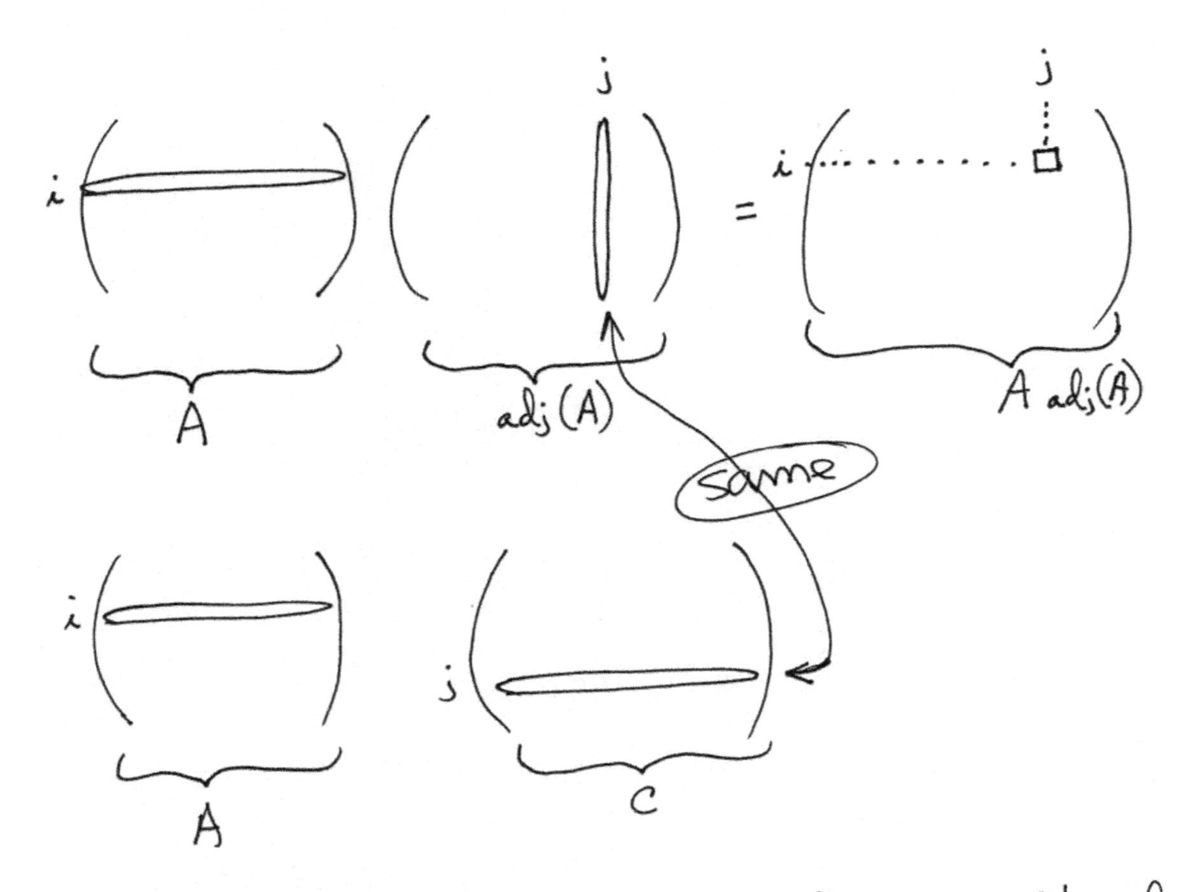

So, the ij-element of $A\ \text{adj}(A)$ is a dot product of the ith row of A and the jth row of C.

We must consider two cases of elements of $A\ \text{adj}(A)$.

Case 1: $i=j$ (diagonal elements)

Then we have a dot product of a row of A with the corresponding row of cofactors. This is exactly the cofactor expansion of $\det A$!

Case 2: $i \neq j$ (non-diagonal elements)

Let's consider a new matrix, A' made by erasing the jth row of A and replacing it with a copy of the ith row.

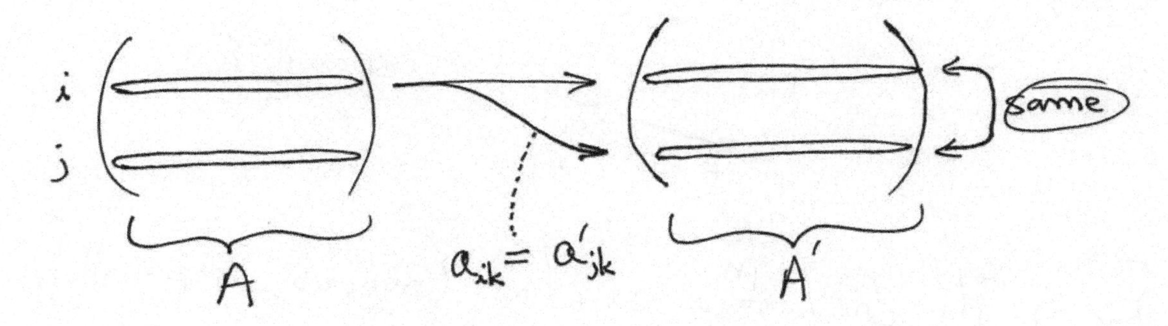

Because A and A' agree <u>off of</u> the jth rows, their cofactor matrices agree <u>on</u> the jth rows.

j — same —

C $\quad$ C'

Of course $\det(A') = 0$; but also, by jth row cofactor exp.:

$$\det(A') = (\text{jth row of } A') \cdot (\text{jth row of } C')$$

$$0!! \quad = (\text{ith row of } A) \cdot (\text{jth row of } C)$$

$$= ij\text{-element of } A \,\text{adj}(A)$$

Linear Independence and Span in $\mathbb{R}^n$

Def: The span of a collection of vectors $\{\vec{v}_1, \ldots, \vec{v}_n\}$ is the set of all linear combinations of those vectors.

Ex: The span of a single (nonzero) vector is a line.

span$\{\vec{v}\}$

$\vec{v}$

$\vec{0}$

Ex: The span of a pair of (nonzero, nonparallel) vectors is a plane.

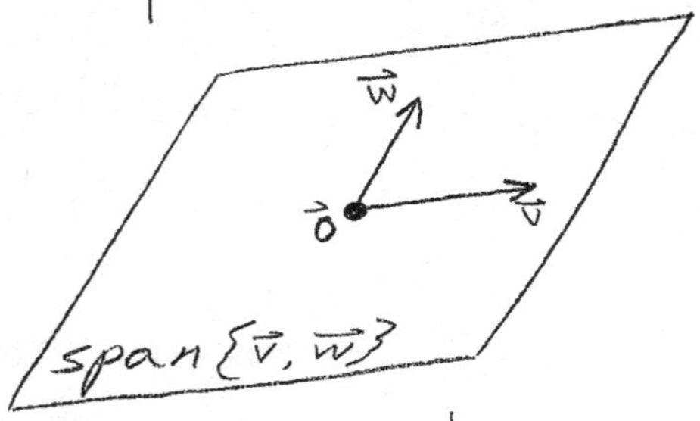

$\vec{w}$

$\vec{0}$ $\vec{v}$

span$\{\vec{v}, \vec{w}\}$

Ex: ... but if the two vectors are parallel, the span is a line.

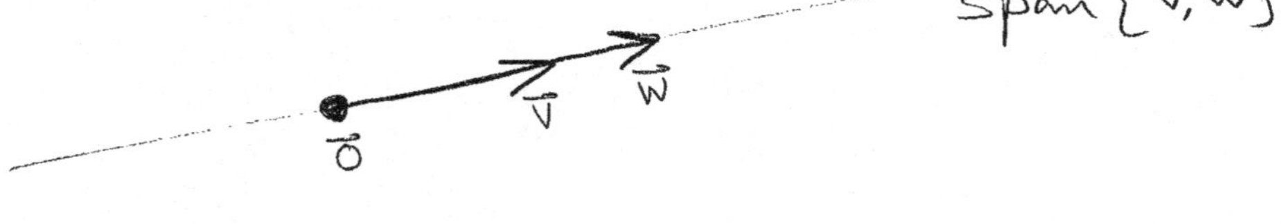

span$\{\vec{v}, \vec{w}\}$

$\vec{v}$ $\vec{w}$

$\vec{0}$

Ex:) The span of $\left\{ \begin{pmatrix} 1 \\ -1 \end{pmatrix}, \begin{pmatrix} 3 \\ 2 \end{pmatrix} \right\}$ would seem to be all of $\mathbb{R}^2$. But how can we be sure?

Need to show that

$$c_1 \begin{pmatrix} 1 \\ -1 \end{pmatrix} + c_2 \begin{pmatrix} 3 \\ 2 \end{pmatrix} = \begin{pmatrix} b_1 \\ b_2 \end{pmatrix}$$

can be solved for all b_1, b_2. Equivalently,

$$\begin{aligned} 1c_1 + 3c_2 &= b_1 \\ -1c_1 + 2c_2 &= b_2 \end{aligned} \quad \text{or} \quad \begin{pmatrix} 1 & 3 \\ -1 & 2 \end{pmatrix} \vec{c} = \vec{b}$$

The matrix is nonsingular, so the system must have a solution.

So the vector equation must have a solution.

So $\text{span}\left\{ \begin{pmatrix} 1 \\ -1 \end{pmatrix}, \begin{pmatrix} 3 \\ 2 \end{pmatrix} \right\} = \mathbb{R}^2$, as desired.

<u>Terminology</u>: We say $\{\vec{v}_1, \ldots, \vec{v}_n\}$ "span" a set $V \subset \mathbb{R}^m$ if $V \subset \text{span}\{\vec{v}_1, \ldots, \vec{v}_n\}$.

Def: A <u>relation</u> for a list of vectors $\{\vec{v}_1, \dots, \vec{v}_n\}$ is a list of coefficients $c_1, \dots, c_n$ such that

$$c_1\vec{v}_1 + \dots + c_n\vec{v}_n = \vec{0}$$

(A relation can also be thought of as a vector, $\vec{c} = \begin{pmatrix} c_1 \\ \vdots \\ c_n \end{pmatrix}$.)

<u>Terminology:</u> The relation $\vec{c} = \vec{0}$ is called the "trivial" relation. Non-trivial relations can also be called "significant".

Def: The list of vectors $\{\vec{v}_1, \dots, \vec{v}_n\}$ is <u>linearly dependent</u> if they have a significant relation.

(Equiv.: $\exists\, c_1, \dots, c_n$, not all zero, with $c_1\vec{v}_1 + \dots + c_n\vec{v}_n = \vec{0}$)

Def: The list of vectors $\{\vec{v}_1, \dots, \vec{v}_n\}$ is <u>linearly independent</u> if the only relation is the trivial relation.

(Equiv.: $(c_1\vec{v}_1 + \dots + c_n\vec{v}_n = \vec{0}) \Rightarrow (c_1, \dots, c_n = 0)$.)

<u>Important!</u> Each of the above is a property of a <u>list</u> of vectors — <u>not</u> of the individual vectors in the list.

So we should say "$\{\vec{v}_1, \dots, \vec{v}_n\}$ <u>is</u> l.d. or l.i.". ← not "are"!

Sloppiness on this is common though...

Here are some examples of linear dependence:

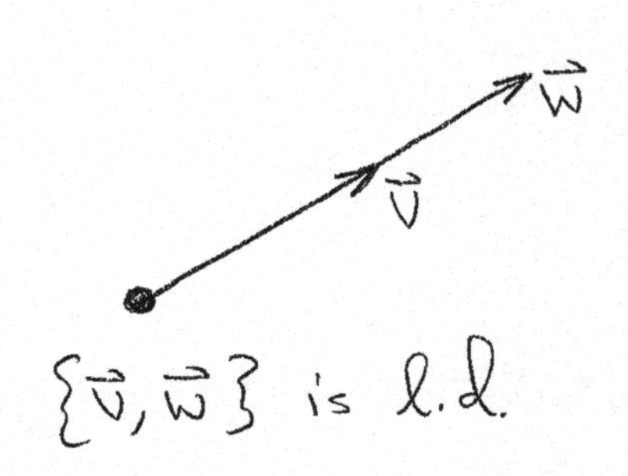

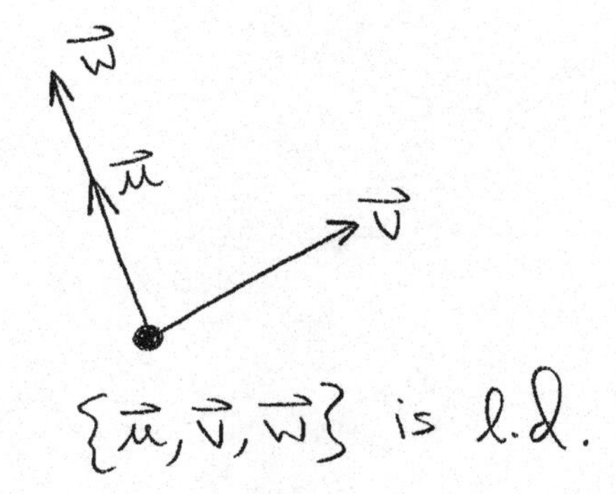

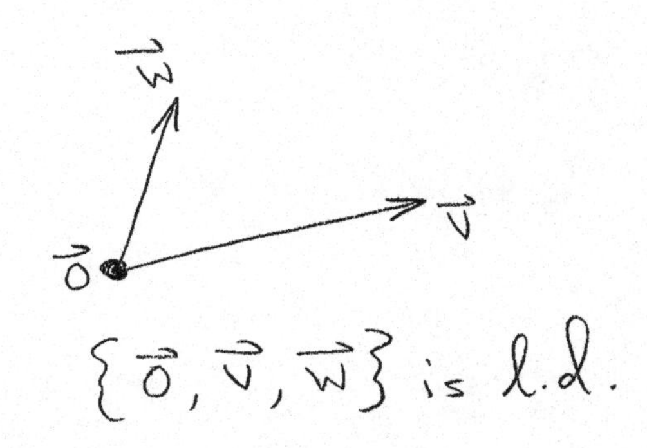

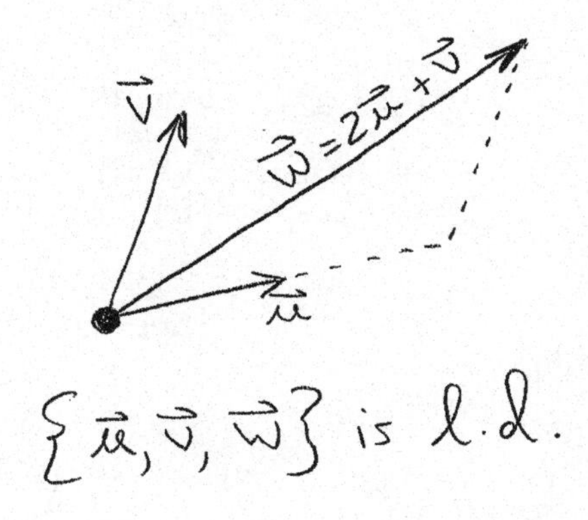

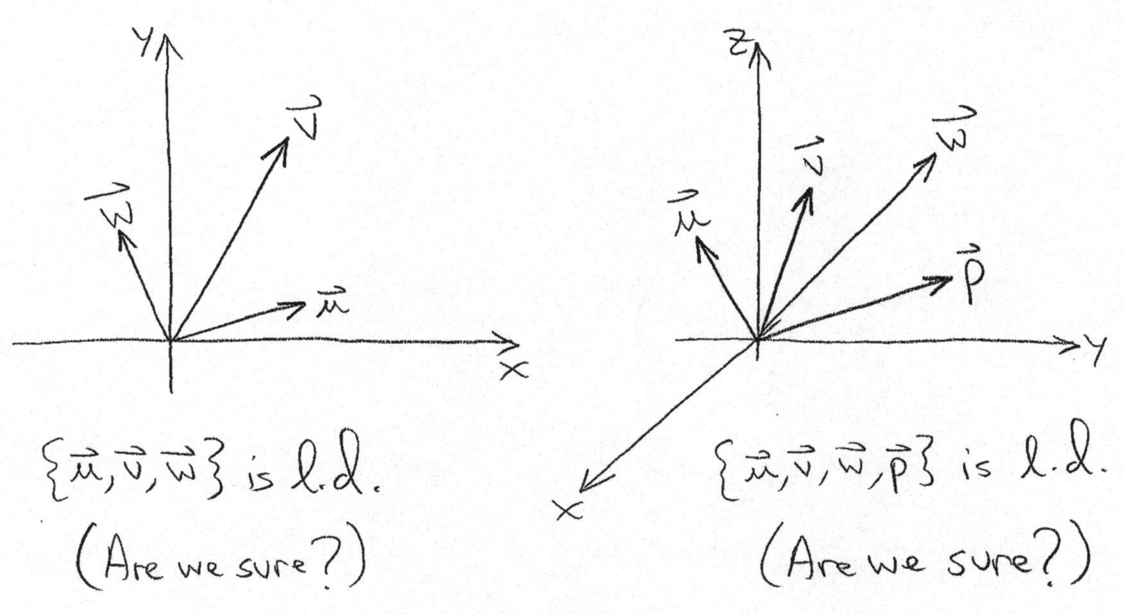

Ex:) Is $\left\{\begin{pmatrix}2\\5\end{pmatrix}, \begin{pmatrix}3\\1\end{pmatrix}\right\}$ l.d. or l.i.?

Consider $\quad c_1\begin{pmatrix}2\\5\end{pmatrix} + c_2\begin{pmatrix}3\\1\end{pmatrix} = \vec{0}$

$$\Leftrightarrow \quad \begin{matrix} 2c_1 + 3c_2 = 0 \\ 5c_1 + 1c_2 = 0 \end{matrix}$$

$$\Leftrightarrow \quad \begin{pmatrix}2 & 3\\5 & 1\end{pmatrix}\vec{c} = \vec{0}$$

The matrix is nonsingular ($\det = -13 \neq 0$), so we have uniqueness, so $\vec{c} = \vec{0}$ is the only sol'n. So the trivial relation is the only relation, and thus the list is l.i..

Thm:) $\left(\{\vec{v}_1, \ldots, \vec{v}_n\} \text{ is l.d.}\right) \Leftrightarrow$ (one of the vectors in the list is a l.c. of the others)

Pf:) ($\Leftarrow$) If $\vec{v}_i = c_1\vec{v}_1 + \ldots \quad \ldots + c_n\vec{v}_n$ (no ith term!)

then $c_1\vec{v}_1 + \ldots + (-1)\vec{v}_i + \ldots + c_n\vec{v}_n = 0$

This relation $\begin{pmatrix}c_1\\ \vdots \\ -1 \\ \vdots \\ c_n\end{pmatrix}$ is significant, so the list is l.d..

($\Rightarrow$) Say $c_1\vec{v}_1 + \dots + c_i\vec{v}_i + \dots + c_n\vec{v}_n = \vec{0}$, with $c_i \neq 0$. Then we can solve for $\vec{v}_i$:

$$\vec{v}_i = \left(-\frac{c_1}{c_i}\right)\vec{v}_1 + \dots \quad \dots + \left(-\frac{c_n}{c_i}\right)\vec{v}_n$$

(no ith term)

So $\vec{v}_i$ is a l.c. of the others.

Note, in the above theorem, the l.c. does not have to be nontrivial!

Ex: $\left\{\begin{pmatrix}0\\0\end{pmatrix}, \begin{pmatrix}1\\0\end{pmatrix}, \begin{pmatrix}0\\1\end{pmatrix}\right\}$ is l.d. b/c

$$1\begin{pmatrix}0\\0\end{pmatrix} + 0\begin{pmatrix}1\\0\end{pmatrix} + 0\begin{pmatrix}0\\1\end{pmatrix} = \begin{pmatrix}0\\0\end{pmatrix}$$

nontrivial!

Alt: $\begin{pmatrix}0\\0\end{pmatrix} = 0\begin{pmatrix}1\\0\end{pmatrix} + 0\begin{pmatrix}0\\1\end{pmatrix}$

trivial!! Okay!

Because there is only one condition to check, the definition is usually easier to work with than this result. But, this result is great for intuition, and sometimes useful in proofs.

Thm: $\left(\{\vec{v}_1, \ldots, \vec{v}_n\} \text{ is l.d.}\right)$

iff

$$\left(\begin{array}{l}\text{there is a vector } \vec{v}_i \text{ such that} \\ \text{span}\{\vec{v}_1, \ldots, \vec{v}_n\} = \text{span}\{\vec{v}_1, \ldots \quad \ldots, \vec{v}_n\} \\ \qquad\qquad\qquad\qquad\qquad \text{(no ith vector)}\end{array}\right)$$

Pf: ($\Rightarrow$) If $\{\vec{v}_1, \ldots, \vec{v}_n\}$ is l.d., then for some $\vec{v}_i$,

$$\vec{v}_i = d_1\vec{v}_1 + \ldots \quad \ldots + d_n\vec{v}_n$$
(no ith term)

Then

$$c_1\vec{v}_1 + \ldots + c_n\vec{v}_n = c_1\vec{v}_1 + \ldots + c_i\left(d_1\vec{v}_1 + \ldots \quad \ldots + d_n\vec{v}_n\right) + \ldots + c_n\vec{v}_n$$
(no ith term)

$$= (c_1 + c_i d_1)\vec{v}_1 + \ldots \quad \ldots + (c_n + c_i d_n)\vec{v}_n$$
(no ith term)

So every l.c. of $\{\vec{v}_1, \ldots, \vec{v}_n\}$ can be rewritten without $\vec{v}_i$. So, $\text{span}\{\vec{v}_1, \ldots, \vec{v}_n\} = \text{span}\{\vec{v}_1, \ldots \quad \ldots, \vec{v}_n\}$
(no ith vector)

($\Leftarrow$) If the spans are equal:

$$\text{span}\{\vec{v}_1, \dots, \vec{v}_n\} = \text{span}\{\vec{v}_1, \dots \quad \dots, \vec{v}_n\}$$

(no ith vector)

(no ith term)

$$\text{Then} \quad v_i \in \text{span}\{\vec{v}_1, \dots, \vec{v}_n\}, \quad v_i = c_1\vec{v}_1 + \dots \quad \dots + c_n\vec{v}_n$$

So $\vec{v}_i$ is a l.c. of the other vectors, and thus $\{\vec{v}_1, \dots, \vec{v}_n\}$ is l.d.

2.1 - Vector Spaces

Here is a theorem about linear combinations of vectors in $\mathbb{R}^n$:

Thm:) A linear combination of linear combinations of vectors in $\mathbb{R}^n$ is a linear combination of those vectors in $\mathbb{R}^n$.

Pf:) Consider the vectors $\vec{v}_1, \ldots, \vec{v}_n$, and their linear combinations

$$\vec{l}_1 = c_{11}\vec{v}_1 + \ldots + c_{1n}\vec{v}_n$$
$$\vec{l}_2 = c_{21}\vec{v}_1 + \ldots + c_{2n}\vec{v}_n$$
$$\vdots$$
$$\vec{l}_k = c_{k1}\vec{v}_1 + \ldots + c_{kn}\vec{v}_n$$

A linear combination of these is

$$\begin{aligned} &d_1\vec{l}_1 + \cdots + d_k\vec{l}_k \\ &= d_1\left(c_{11}\vec{v}_1 + \cdots + c_{1n}\vec{v}_n\right) + \ldots + d_k\left(c_{k1}\vec{v}_1 + \ldots + c_{kn}\vec{v}_n\right) \\ &= \left(d_1c_{11} + \ldots + d_kc_{k1}\right)\vec{v}_1 + \ldots + \left(d_1c_{1n} + \ldots + d_kc_{kn}\right)\vec{v}_n \end{aligned}$$

which is therefore a linear combination of $\vec{v}_1, \ldots, \vec{v}_n$, as required.

Suppose we later find ourselves interested in taking linear combinations of functions... The ideas in the previous proof seem to make it clear that an analogous theorem and proof would work — but we have to state/prove separately because the previous only applies to vectors in $\mathbb{R}^n$.

Thm:) A linear combination of linear combinations of functions is a linear combination of those functions.

Pf:) Consider the functions $f_1, \ldots, f_n$, and their linear combinations

$$g_1 = c_{11}f_1 + \ldots + c_{1n}f_n$$
$$g_2 = c_{21}f_1 + \ldots + c_{2n}f_n$$
$$\vdots$$
$$g_k = c_{k1}f_1 + \ldots + c_{kn}f_n$$

A linear combination of these is

$$d_1 g_1 + \ldots + d_k g_k$$
$$= d_1(c_{11}f_1 + \ldots + c_{1n}f_n) + \ldots + d_k(c_{k1}g_1 + \ldots + c_{kn}f_n)$$
$$= (d_1 c_{11} + \ldots + d_k c_{k1})f_1 + \ldots + (d_1 c_{1n} + \ldots + d_k c_{kn})f_n$$

And then what if you later need to understand linear combinations of other kinds of things?

You would have to state and prove yet another theorem, seemingly the same as the two previous.... ☹

This is tedious!!

<u>Solution</u>: Define an <u>abstraction</u>, in which you do not specify what the objects <u>are</u> — only the <u>properties</u> that they satisfy.

Choose properties that you <u>need</u> for what you are doing.

State/prove the theorem about the <u>abstraction</u>, and then it automatically applies to specific examples.

For the previous theorems we need to be able to:

- add objects
- multiply objects by scalars
- do "standard" algebra of these operations

The abstraction we define with this in mind is called a "vector space over $\mathbb{R}$".

<u>Def:</u> A <u>vector space over $\mathbb{R}$</u> is a trio of

(a) a set V, whose elements are called "vectors"

(b) an operation of "addition" of the vectors in V

(c) an operation of "scalar multiplication" of the vectors in V by real scalars

satisfying these eight properties:

(1) $u + v = v + u$ for all $u, v \in V$

(2) $u + (v + w) = (u + v) + w$ ....

(3) There is a "zero vector", 0, with

$$0 + v = v \quad \text{for all } v \in V$$

(4) For each $v \in V$ there is an "additive inverse", $(-v)$, with

$$v + (-v) = 0$$

(5) $c(u+v) = cu + cv$ for all $u, v \in V$, $c \in \mathbb{R}$

(6) $c(dv) = (cd)v$ for all $v \in V$, $c, d \in \mathbb{R}$

(7) $(c+d)v = cv + dv$ ....

(8) $1 \cdot v = v$ for all $v \in V$

("over $\mathbb{R}$" refers to the scalars being real. We will later consider vector spaces with different scalars.)

Ex: $\mathbb{R}^n$ is a vector space, with the usual operations. (Check!)

Ex: $F(a,b)$ is the set of real-valued functions on the interval $(a,b) \subset \mathbb{R}$, with

ⓑ "+" defined by (vector addition)

$(f+g)(x) = f(x) + g(x)$ (real addition)

ⓒ scalar product defined by (scalar mult.)

$(cf)(x) = cf(x)$ (real mult.)

This satisfies all eight properties, so (check!) $F(a,b)$ is a vector space over $\mathbb{R}$.

① $(f+g)(x) = f(x) + g(x)$
$(g+f)(x) = g(x) + f(x)$
(these are equal b/c real addition is commutative!)

So $f+g$ and $g+f$ always have the same values, and thus they are the same function.

Now let's state and prove the previous theorem in the abstract:

Thm:) A linear combination of linear combinations of vectors in a vector space V is a linear combination of those vectors.

Pf:) Consider $v_1, \dots, v_n \in V$ and

$$l_1 = c_{11} v_1 + \dots + c_{1n} v_n$$
$$\vdots$$
$$l_k = c_{k1} v_1 + \dots + c_{kn} v_n$$

A linear combination of these is

$$\begin{aligned} d_1 l_1 + \dots + d_k l_k &= d_1 (c_{11} v_1 + \dots + c_{1n} v_n) + \dots + d_k (c_{k1} v_1 + \dots + c_{kn} v_n) \\ &= (d_1 c_{11} + \dots + d_k c_{k1}) v_1 + \dots + (d_1 c_{1n} + \dots + d_k c_{kn}) v_n \end{aligned}$$

which is a linear combination of $v_1, \dots, v_n$ as required.

We have already noted that $\mathbb{R}^n$, $F(a,b)$ are vector spaces, so <u>both</u> of the previously stated theorems follow immediately from this one.

Ex:) $M_{3\times 5}(\mathbb{R})$ is the set of all 3×5 matrices with real entries; and we have an addition and a scalar multiplication.

Check that the eight properties are satisfied, and we can conclude that $M_{3\times 5}(\mathbb{R})$ is a vector space.

And then it immediately follows that the abstract theorem applies to these matrices!

Not everything is a vector space, of course...

Ex:) $W = \{\vec{x} \in \mathbb{R}^3 \mid x+y+z=1\}$ is <u>not</u> a vector space with the usual addition.

In fact, the usual vector addition is not even an operation on W.

$$\begin{pmatrix}1\\0\\0\end{pmatrix} + \begin{pmatrix}0\\1\\0\end{pmatrix} = \begin{pmatrix}1\\1\\0\end{pmatrix}$$

(these are in W) (this is <u>not</u> in W)

Ex: $V = \{ \vec{x} \in \mathbb{R}^3 \mid x+y+z=1 \}$, with operations defined by:

$$\begin{pmatrix} x_1 \\ y_1 \\ z_1 \end{pmatrix} + \begin{pmatrix} x_2 \\ y_2 \\ z_2 \end{pmatrix} = \begin{pmatrix} x_1+x_2-1 \\ y_1+y_2 \\ z_1+z_2 \end{pmatrix}$$

$$c\begin{pmatrix} x \\ y \\ z \end{pmatrix} = \begin{pmatrix} cx-c+1 \\ cy \\ cz \end{pmatrix}$$

is a vector space.

$\begin{pmatrix} 1 \\ 0 \\ 0 \end{pmatrix}$ is the "0" vector

$\begin{pmatrix} 2-x \\ -y \\ -z \end{pmatrix}$ is the "$-\vec{x}$" vector

(Check other conditions!)

Facts:
1. The zero vector of V is unique.
2. The negative of a vector $v \in V$ is unique.
3. $0 \cdot \vec{v} = \vec{0}$
4. $c \cdot \vec{0} = \vec{0}$
5. $(-1)\vec{v} = -\vec{v}$

2.2 - Subspaces

Def: $W \subset V$ is a <u>subspace</u> of V if W, using the same operations as V, is a vector space.

Ex: $\left\{ \begin{pmatrix} x \\ y \\ 0 \end{pmatrix} \middle| \ x, y \in \mathbb{R} \right\}$ is a subspace of $\mathbb{R}^3$. (Check)

Note that in order for W to be a subspace of V, we need for the operations on V to be operations on W.

That is, we need

$$\vec{w}_1, \vec{w}_2 \in W \implies \vec{w}_1 + \vec{w}_2 \in W$$
$$\vec{w} \in W \implies c\vec{w} \in W$$

<u>Terminology</u>: W must be "closed under addition" and "closed under scalar multiplication".

It turns out that this is enough!

Thm: If W is a nonempty subset of V that is closed under addition and scalar multiplication, then W is a subspace of V.

Pf:) Most of the conditions are on the operations, and they are already satisfied because V is a vector space!

We just need to check two remaining properties:

③ Take any $\vec{w} \in W$, and recall that

$$\vec{0} = 0\vec{w}$$

Since we are given W is closed under scalar multiplication, we have $\vec{0} \in W$.

④ For any $\vec{w} \in W$, recall also that

$$(-\vec{w}) = (-1)\vec{w}$$

Since we are given W is closed under scalar multiplication, we have $(-\vec{w}) \in W$.

Ex: Consider $W \subset F(a,b)$ defined by

$$W = \left\{ h \in F(a,b) \;\middle|\; h(m) = 0 \,, \ (\text{where } m = \tfrac{a+b}{2}) \right\}$$

Is W closed under addition?

Suppose $f, g \in W$.

Then $f(m) = 0$, and $g(m) = 0$.

We compute $(f+g)(m)$ by

$(f+g)(m) = f(m) + g(m) = 0 + 0 = 0$.

So $f+g \in W$, and it follows that W is closed under addition.

Is W closed under scalar multiplication?

Suppose $f \in W$.

Then $f(m) = 0$.

We compute $(cf)(m)$ by

$(cf)(m) = c\,f(m) = c \cdot 0 = 0$.

So $cf \in W$, and it follows that W is closed under scalar multiplication.

Ex:) Let $C(a,b) = \{\text{continuous fns on } (a,b)\}$

$D(a,b) = \{\text{diff'bl fns on } (a,b)\}$

We know every diff'bl fn is continuous; and $D(a,b)$ is closed under addition and scalar mult.

So $D(a,b)$ is a subspace of $C(a,b)$

Ex:) $D^n(a,b) = \{f \mid f \text{ has an nth deriv.}\}$

$C^n(a,b) = \{f \mid f^{[n]} \text{ is continuous}\}$

These are all subspaces of $F(a,b)$, and

$$C = C^0 \supset D^1 \supset C^1 \supset D^2 \supset C^2 \supset \cdots$$

Ex:) $C^\infty = \{f \mid f \text{ has } \infty\text{'ly many derivatives}\}$ is a subspace of <u>all</u> of the above.

Ex:) $P = \{\text{all polynomials}\}$ is a subspace of C^∞.

Ex:) For any set of vectors $v_1, \ldots, v_n \in V$,

$$\text{span}\{v_1, \ldots, v_n\}$$

is a subspace of V. (Check!)

Ex:) For any $m \times n$ matrix A, the set of homogeneous solutions

$$H = \{\vec{x} \in \mathbb{R}^n \mid A\vec{x} = \vec{0}\}$$

is a subspace of $\mathbb{R}^n$.

Check:

If $\vec{x}_1, \vec{x}_2 \in H$ (in which case $A\vec{x}_1 = \vec{0}$, $A\vec{x}_2 = \vec{0}$),

then $A(\vec{x}_1 + \vec{x}_2) = A\vec{x}_1 + A\vec{x}_2 = \vec{0} + \vec{0} = \vec{0}$,

so $\vec{x}_1 + \vec{x}_2 \in H$.

If $\vec{x} \in H$ (in which case $A\vec{x} = \vec{0}$),

then $A(c\vec{x}) = c(A\vec{x}) = c\vec{0} = \vec{0}$,

so $c\vec{x} \in H$.

So H is closed under addition and scalar multiplication, and thus is a subspace.

Linear Independence, Span, and Bases in Vector Spaces

We can generalize our previous ideas of span and linear independence to vector spaces.

Def: The span of a collection $\{v_1, \ldots, v_n\}$ of vectors in V is the set of all linear combinations of those vectors.

Def: The list $\{v_1, \ldots, v_n\}$ of vectors in V is linearly dependent if they have a significant relation.

(Equiv: $\exists\, c_1, \ldots, c_n$, not all zero, with $c_1 v_1 + \ldots + c_n v_n = 0$)

Def: The list $\{v_1, \ldots, v_n\}$ of vectors in V is linearly independent if the only relation is the trivial relation.

(Equiv.: $(c_1 v_1 + \ldots + c_n v_n = 0) \Longrightarrow (c_1, \ldots, c_n = 0)$)

Ex:) Consider $p_1, p_2 \in P$, with $p_1(x) = x^2 - 1$, $p_2(x) = x + 5$.

Is $\{p_1, p_2\}$ linearly independent?

Suppose $c_1 p_1 + c_2 p_2 = 0$ (the zero function $\in P$)

Then $c_1(x^2 - 1) + c_2(x + 5) = 0$

$$(c_1)x^2 + (c_2)x + (-c_1 + 5c_2)1 = 0$$

All coefficients must be zero, so

$$\begin{aligned} c_1 \quad\quad &= 0 \\ c_2 &= 0 \\ -c_1 + 5c_2 &= 0 \end{aligned}$$

$\Rightarrow c_1, c_2 = 0$

So the only relation is the trivial relation, and thus $\{p_1, p_2\}$ is linearly independent.

Def:) The list $\{v_1, \ldots, v_n\}$ of vectors in V is a <u>basis</u> for V if both:

① $\{v_1, \ldots, v_n\}$ is linearly independent

<u>and</u>

② $\{v_1, \ldots, v_n\}$ spans V

Because they span V, we can say that every vector in V can be represented by the vectors in $\{\vec{v}_1, \dots, \vec{v}_n\}$

Because they are l.i., we cannot remove any of these vectors and still span. So, none of the vectors are "unnecessary".

Ex: Are $\left\{ \begin{pmatrix} 3 \\ 2 \end{pmatrix}, \begin{pmatrix} 1 \\ 5 \end{pmatrix} \right\}$ a basis for $\mathbb{R}^2$?

To see that they span, we need existence in

$$c_1 \begin{pmatrix} 3 \\ 2 \end{pmatrix} + c_2 \begin{pmatrix} 1 \\ 5 \end{pmatrix} = \begin{pmatrix} b_1 \\ b_2 \end{pmatrix} \iff \begin{pmatrix} 3 & 1 \\ 2 & 5 \end{pmatrix} \begin{pmatrix} c_1 \\ c_2 \end{pmatrix} = \begin{pmatrix} b_1 \\ b_2 \end{pmatrix}$$

To see that they are l.i., we need uniqueness in

$$c_1 \begin{pmatrix} 3 \\ 2 \end{pmatrix} + c_2 \begin{pmatrix} 1 \\ 5 \end{pmatrix} = \begin{pmatrix} 0 \\ 0 \end{pmatrix} \iff \begin{pmatrix} 3 & 1 \\ 2 & 5 \end{pmatrix} \begin{pmatrix} c_1 \\ c_2 \end{pmatrix} = \begin{pmatrix} 0 \\ 0 \end{pmatrix}$$

The matrix is nonsingular, so we have both conditions.

Ex:) The "standard basis vectors" in $\mathbb{R}^n$

$$\left\{ \begin{pmatrix} 1 \\ 0 \\ \vdots \\ 0 \end{pmatrix}, \begin{pmatrix} 0 \\ 1 \\ 0 \\ \vdots \\ 0 \end{pmatrix}, \dots, \begin{pmatrix} 0 \\ \vdots \\ 0 \\ 1 \end{pmatrix} \right\}$$

are a basis for $\mathbb{R}^n$.

Similarly to previous example, this is because

$$\begin{pmatrix} 1 & & 0 \\ & \ddots & \\ 0 & & 1 \end{pmatrix}$$

is nonsingular.

<u>Notation reminder</u>:

$$\vec{e}_i = \begin{pmatrix} 0 \\ \vdots \\ 0 \\ 1 \\ 0 \\ \vdots \\ 0 \end{pmatrix} \leftarrow 1 \text{ is in the } i\text{th entry}$$

Ex:) Are $\left\{ \begin{pmatrix} 1 \\ 2 \\ -3 \end{pmatrix}, \begin{pmatrix} 1 \\ 1 \\ -2 \end{pmatrix} \right\}$ a basis for $\left\{ \vec{x} \mid x+y+z=0 \right\}$?

For l.i., need uniqueness in $\begin{pmatrix} 1 & 1 \\ 2 & 1 \\ -3 & -2 \end{pmatrix}$. This works. (Check!)

For span, we do <u>not</u> need a pivot in every row though, because not all $\vec{b}$ are required...

Option 1: Check for existence in

$$c_1\begin{pmatrix}1\\2\\-3\end{pmatrix} + c_2\begin{pmatrix}1\\1\\-2\end{pmatrix} = \underbrace{\begin{pmatrix}b_1\\b_2\\-b_1-b_2\end{pmatrix}}$$

These are all of the vectors in $\{x+y+z=0\}$

$$\left(\begin{array}{cc|c}1 & 1 & b_1\\2 & 1 & b_2\\-3 & -2 & -b_1-b_2\end{array}\right)$$

Option 2: Note solutions to $x+y+z=0$ are

$$\begin{pmatrix}-y-z\\y\\z\end{pmatrix} = y\begin{pmatrix}-1\\1\\0\end{pmatrix} + z\begin{pmatrix}-1\\0\\1\end{pmatrix}$$

So $\begin{pmatrix}-1\\1\\0\end{pmatrix}, \begin{pmatrix}-1\\0\\1\end{pmatrix}$ span the space.

If these are l.c.'s of $\begin{pmatrix}1\\2\\-3\end{pmatrix}, \begin{pmatrix}1\\1\\-2\end{pmatrix}$, then these span the space too.

So, need to check for existence in

$$\left(\begin{array}{cc|c}1 & 1 & -1\\2 & 1 & 1\\-3 & -2 & 0\end{array}\right) \text{ and } \left(\begin{array}{cc|c}1 & 1 & -1\\2 & 1 & 0\\-3 & -2 & 1\end{array}\right)$$

We will soon find an easier way to answer these kinds of questions.

Thm:) $\{\vec{v}_1, \ldots, \vec{v}_n\}$ in V are a basis for V

iff

each vector in V is uniquely expressible as a l.c. of $\{\vec{v}_1, \ldots, \vec{v}_n\}$ in V.

Pf:) ($\Rightarrow$) We assume $\{\vec{v}_1, \ldots, \vec{v}_n\}$ is a basis for V, so we know we have span and l.i.

Because they span V, we know we can write

$$\vec{v} = c_1\vec{v}_1 + \ldots + c_n\vec{v}_n$$

for any $\vec{v} \in V$.

If we can also write

$$\vec{v} = d_1\vec{v}_1 + \ldots + d_n\vec{v}_n$$

then

$$\vec{v} - \vec{v} = (c_1 - d_1)\vec{v}_1 + \ldots + (c_n - d_n)\vec{v}_n = \vec{0}$$

Because of l.i., this means $c_1 = d_1, \ldots, c_n = d_n$.

So the expression is unique.

($\Leftarrow$) If every vector is uniquely expressible, then certainly every vector is expressible, so $\{\vec{v}_1, \ldots, \vec{v}_n\}$ span V.

By uniqueness, we have

$$c_1\vec{v}_1 + \ldots + c_n\vec{v}_n = \vec{0} \implies c_1 = 0, \ldots, c_n = 0.$$

So the vectors are l.i., and thus are a basis.

So a basis is not only an "efficient" way to represent elements of a vector space (can't remove vectors from a basis), it is also a way to do so unambiguously

Def: Say $\alpha = \{\vec{v}_1, \ldots, \vec{v}_n\}$ is a basis for V, and $\vec{v} \in V$. Then the coordinates of $\vec{v}$ relative to the basis α are

$$[\vec{v}]_\alpha = \begin{pmatrix} c_1 \\ \vdots \\ c_n \end{pmatrix}$$

where $\vec{v} = c_1\vec{v}_1 + \ldots + c_n\vec{v}_n$

Note: What we usually refer to as the "coordinates" of a vector $\vec{v}$ are the "coordinates relative to the standard basis, because, for example,

$$\begin{pmatrix} a \\ b \end{pmatrix} = a\begin{pmatrix} 1 \\ 0 \end{pmatrix} + b\begin{pmatrix} 0 \\ 1 \end{pmatrix} = a\vec{e}_1 + b\vec{e}_2$$

Ex:) Let β be the basis $\{x^2-x+1,\ x-1,\ 1\}$ for the vector space of polynomials of degree ≤ 2.

Say f is the vector $f(x) = 3x^2 + 2x + 5$.

What is $[f]_\beta$?

We need to write

$$c_1(x^2-x+1) + c_2(x-1) + c_3(1) = 3x^2+2x+5$$

$$\begin{aligned} c_1 - c_2 + c_3 &= 5 && \text{(const. terms)} \\ -c_1 + c_2 &= 2 && (x \text{ terms}) \\ c_1 &= 3 && (x^2 \text{ terms}) \end{aligned}$$

The unique solution is $c_1 = 3,\ c_2 = 5,\ c_3 = 7$

So we have

$$[f]_\beta = \begin{pmatrix} 3 \\ 5 \\ 7 \end{pmatrix}$$

2.4 – Dimension & Matrix Subspaces

Dimension

Thm: If $\{\vec{v}_1, \ldots, \vec{v}_n\}$ is a basis for a v.s. V, then any set $\{\vec{w}_1, \ldots, \vec{w}_m\}$ of $m > n$ vectors is l.d.

Pf: Since $\{\vec{v}_1, \ldots, \vec{v}_n\}$ is a basis, we can write each $\vec{w}_i$ as a l.c.:

$$\begin{array}{cccc}
\vec{w}_1 & \vec{w}_2 & & \vec{w}_m \\
\| & \| & \cdots & \| \\
a_{11}\vec{v}_1 & a_{12}\vec{v}_1 & & a_{1m}\vec{v}_1 \\
+ & + & & + \\
a_{21}\vec{v}_2 & a_{22}\vec{v}_2 & & a_{2m}\vec{v}_2 \\
+ & + & & + \\
\vdots & \vdots & & \vdots \\
+ & + & & + \\
a_{n1}\vec{v}_n & a_{n2}\vec{v}_n & & a_{nm}\vec{v}_n
\end{array}$$

Let A be the matrix of these coefficients:

$$A = \begin{pmatrix} a_{11} & a_{12} & \cdots & a_{1m} \\ a_{21} & a_{22} & \cdots & a_{2m} \\ \vdots & \vdots & & \vdots \\ a_{n1} & a_{n2} & \cdots & a_{nm} \end{pmatrix}$$

<u>Lemma</u>:
$$c_1\vec{w}_1 + \cdots + c_m\vec{w}_m = d_1\vec{v}_1 + \cdots + d_n\vec{v}_n$$
iff
$$A\vec{c} = \vec{d}$$

<u>Pf of Lemma</u>:
$$c_1\vec{w}_1 + \cdots + c_m\vec{w}_m$$
$$= c_1\begin{pmatrix} a_{11}\vec{v}_1 \\ + \\ a_{21}\vec{v}_2 \\ + \\ \vdots \\ + \\ a_{n1}\vec{v}_n \end{pmatrix} + c_2\begin{pmatrix} a_{12}\vec{v}_1 \\ + \\ a_{22}\vec{v}_2 \\ + \\ \vdots \\ + \\ a_{n2}\vec{v}_n \end{pmatrix} + \cdots + c_m\begin{pmatrix} a_{1m}\vec{v}_1 \\ + \\ a_{2m}\vec{v}_2 \\ + \\ \vdots \\ + \\ a_{nm}\vec{v}_n \end{pmatrix}$$
$$\begin{aligned} &= (c_1a_{11} + c_2a_{12} + \cdots + c_ma_{1m})\vec{v}_1 \\ &+ (c_1a_{21} + c_2a_{22} + \cdots + c_ma_{2m})\vec{v}_2 \\ &\;\vdots \\ &+ (c_1a_{n1} + c_2a_{n2} + \cdots + c_ma_{nm})\vec{v}_n \end{aligned} = \begin{matrix} d_1\vec{v}_1 \\ + \\ d_2\vec{v}_2 \\ + \\ \vdots \\ + \\ d_n\vec{v}_n \end{matrix}$$

So
$$\begin{aligned} a_{11}c_1 + a_{12}c_2 + \cdots + a_{1m}c_m &= d_1 \\ a_{21}c_1 + a_{22}c_2 + \cdots + a_{2m}c_m &= d_2 \\ &\;\vdots \\ a_{n1}c_1 + a_{n2}c_2 + \cdots + a_{nm}c_m &= d_n \end{aligned}$$

and thus $A\vec{c} = \vec{d}$

(back to main proof...)

If $m > n$, then A has more columns than rows. So,

$$A\vec{c} = \vec{0}$$

has a nontrivial solution. By the lemma, this means there are nontrivial $c_1, \ldots, c_m$ with

$$c_1\vec{w}_1 + \cdots + c_m\vec{w}_m = 0\vec{v}_1 + \cdots + 0\vec{v}_n = \vec{0}$$

So $\{\vec{w}_1, \ldots, \vec{w}_m\}$ is l.d. ∎

This theorem allows us to prove another important result.

Thm: If $\{\vec{v}_1, \ldots, \vec{v}_n\}$, $\{\vec{u}_1, \ldots, \vec{u}_m\}$ are both bases for V, then $n = m$.

Pf: If $m > n$, the previous thm implies $\{\vec{u}_1, \ldots, \vec{u}_m\}$ l.d. ✗

If $n > m$, ... $\{\vec{v}_1, \ldots, \vec{v}_n\}$ l.d. ✗

So we must have $n = m$.

So for any vector space, the number of elements in a basis is the same for every basis.

That is — this number is an <u>invariant</u>. It is intrinsic to the space V, not just an individual basis.

Def: If V has a basis with n elements, then we say the <u>dimension</u> of V is n, and write

$$\dim(V) = n$$

Ex: $\dim(\mathbb{R}^n) = n$ (consider standard basis).

Ex: $\dim(M_{m \times n}) = mn$ (find a basis!)

Nontrivial vector spaces with no finite basis are called "infinite dimensional".

Ex: $C^0(a,b)$ is infinite dimensional.

Facts proved in the book:

1. If $\{v_1, \ldots, v_k\}$ is l.i. in V, with $\dim(V) = n$, then one can find $v_{k+1}, \ldots, v_n \in V$ such that $\{v_1, \ldots, v_n\}$ is a basis for V.

2. If $\{v_1, \ldots, v_k\}$ span V, then there is a subset that is a basis for V.

3. If $\dim(V) = n$ and $\{v_1, \ldots, v_n\}$ in V is l.i., then it is a basis.

4. If $\dim(V) = n$ and $\{v_1, \ldots, v_n\}$ in V spans V, then it is a basis.

Ex: It is easy to check that $\{1, x, x^2\}$ is a basis for P_2 (check!). Suppose we have also confirmed that $\{x^2 - 2x + 7,\ x+1,\ 3\}$ is l.i.

Then, by (3) above, we can conclude it is a basis for P_2 (and thus spans P_2) without directly checking the span.

Nullspace

(sometimes called the "kernel")

Def: The nullspace of a matrix A is the set of homogeneous solutions

$$NS(A) = \{ \vec{x} \mid A\vec{x} = \vec{0} \}$$

We already know how to solve systems; but in fact this method also gives us a basis for NS.

Ex: Suppose A has rref given by

$$\text{rref}(A) = \begin{pmatrix} 1 & 3 & 0 \\ 0 & 0 & 1 \end{pmatrix}$$

Then the homogeneous solutions are

$$\vec{x} = \begin{pmatrix} x \\ y \\ z \end{pmatrix} = \begin{pmatrix} -3y \\ y \\ 0 \end{pmatrix} = y \begin{pmatrix} -3 \\ 1 \\ 0 \end{pmatrix}$$

This is a 1-dim. sol. set, with basis $\left\{ \begin{pmatrix} -3 \\ 1 \\ 0 \end{pmatrix} \right\}$

Ex:) Suppose A reduces to

$$\text{rref}(A) = \begin{pmatrix} 1 & 4 & 0 & 2 \\ 0 & 0 & 1 & 3 \end{pmatrix}$$

The homogeneous solutions are given by

$$\vec{x} = \begin{pmatrix} x \\ y \\ z \\ w \end{pmatrix} = \begin{pmatrix} -4y-2w \\ y \\ -3w \\ w \end{pmatrix} = y\begin{pmatrix} -4 \\ 1 \\ 0 \\ 0 \end{pmatrix} + w\begin{pmatrix} -2 \\ 0 \\ -3 \\ 1 \end{pmatrix}$$

This is a 2-dim set, with basis given by these.

How do we know these are independent?
A relation would be represented by y, w with

$$y\begin{pmatrix} -4 \\ 1 \\ 0 \\ 0 \end{pmatrix} + w\begin{pmatrix} -2 \\ 0 \\ -3 \\ 1 \end{pmatrix} = \begin{pmatrix} 0 \\ 0 \\ 0 \\ 0 \end{pmatrix}$$

whose 2nd and 4th equations clearly show $y, w = 0$.

Thm:) $\dim(NS(A)) = \#$ free variables in $\text{rref}(A)$.

Row Space

Def: The <u>row space</u> of A is the span of the row vectors of A. It is written $RS(A)$.

Thm: If E is an elementary matrix, then

$$RS(A) = RS(EA)$$

Pf: Every row of each matrix is a l.c. of rows of the other matrix, so their spans are equal.

Given the above thm, we can also conclude

$$RS(A) = RS\big(\text{rref}(A)\big)$$

Conveniently, the non zero rows of rref(A) are a basis for $RS(\text{rref}(A))$. So, they are also a basis for $RS(A)$.

Ex:

$$A = \begin{pmatrix} 1 & 2 & -1 & 3 & 0 \\ 1 & 1 & 0 & 4 & 1 \\ 1 & 4 & -3 & 1 & -2 \end{pmatrix} \Rightarrow \text{rref}(A) = \begin{pmatrix} 1 & 0 & +1 & 5 & 2 \\ 0 & 1 & -1 & -1 & -1 \\ 0 & 0 & 0 & 0 & 0 \end{pmatrix}$$

Clearly $\left\{ \begin{pmatrix} 1 \\ 0 \\ 1 \\ 5 \\ 2 \end{pmatrix}, \begin{pmatrix} 0 \\ 1 \\ -1 \\ -1 \\ -1 \end{pmatrix} \right\}$ is a basis for $RS(\text{rref}(A))$,

so they are also for $RS(A)$.

Note, we have a basis vector for $RS(A)$ for every pivot...

And we had a basis vector for $NS(A)$ for every free variable...

So, $\dim(NS(A)) + \dim(RS(A)) = \#$ cols of A

We also have the following:

Thm: $RS(A) \perp NS(A)$

Pf: If $\vec{r} \in RS(A)$, then $\vec{r} = c_1\vec{r}_1 + \cdots + c_k\vec{r}_k$

If $\vec{n} \in NS(A)$, then $A\vec{n} = \vec{0}$, so $\vec{r}_i \cdot \vec{n} = 0$ for all i.

Then we can compute $\vec{r} \cdot \vec{n}$ by

$$\begin{aligned}\vec{r}\cdot\vec{n} &= (c_1\vec{r}_1 + \cdots + c_k\vec{r}_k)\cdot\vec{n} \\ &= c_1(\vec{r}_1\cdot\vec{n}) + \cdots + c_k(\vec{r}_k\cdot\vec{n}) \\ &= 0\end{aligned}$$

So every vector in $RS(A)$ is $\perp$ to every vector in $NS(A)$.

It can also be shown that

① $NS(A) = RS(A)^{\perp}$ (the set of all vectors $\perp$ to the row space.)

② $RS(A) = NS(A)^{\perp}$ (the set of all vectors $\perp$ to the null space.)

③ If $\{\vec{v}_1, \ldots, \vec{v}_k\}$ is a basis for $RS(A)$ and $\{\vec{w}_1, \ldots, \vec{w}_{n-k}\}$ is a basis for $NS(A)$, then $\{\vec{v}_1, \ldots, \vec{v}_k, \vec{w}_1, \ldots, \vec{w}_{n-k}\}$ is a basis for $\mathbb{R}^n$.

Column Space

Def:) The column space of A is the span of the column vectors of A. It is written $CS(A)$.

Row operations <u>do not</u> preserve the column space; They <u>do</u> preserve <u>relations between column vectors.</u>

Ex:)

$$\begin{pmatrix} 1 & 3 \\ 2 & 6 \end{pmatrix} \quad \leftarrow \quad 3\vec{c}_1 - 1\vec{c}_2 = \vec{0}$$

$$\begin{pmatrix} 1 & 3 \\ 3 & 9 \end{pmatrix} \begin{matrix} \textcircled{1} \\ \textcircled{2} + \textcircled{1} \end{matrix} \quad \leftarrow \quad 3\vec{c}_1{}' - 1\vec{c}_2{}' = \vec{0}$$

Thm:) If E is an elementary matrix, then a relation between the columns of A is also a relation between the columns of EA.

Pf:) Recalling that a matrix-vector product is a l.c. of columns, we can write the relations in question as

$$A\vec{c} = \vec{0} \quad \text{and} \quad (EA)\vec{c} = \vec{0}$$

These statements are equivalent because E is invertible.

Note, the pivot columns of rref(A) are a basis for CS(rref(A)).

Ex:) $\begin{pmatrix} 1 & 0 & 1 & 5 & 2 \\ 0 & 1 & -1 & -1 & -1 \\ 0 & 0 & 0 & 0 & 0 \end{pmatrix} = \text{rref}(A)$

$\uparrow$ under each column: $\vec{v}_1 \quad \vec{v}_2 \quad \vec{v}_3 \quad \vec{v}_4 \quad \vec{v}_5$

$\{\vec{v}_1, \vec{v}_2\}$ is a basis for CS(rref(A)) because:

① they are independent because

$$c_1\vec{v}_1 + c_2\vec{v}_2 = \vec{0}$$

has only the trivial solution

② they span CS(rref(A)) because

$\vec{v}_3 = d_{31}\vec{v}_1 + d_{32}\vec{v}_2$

$\vec{v}_4 = d_{41}\vec{v}_1 + d_{42}\vec{v}_2$

$\vec{v}_5 = d_{51}\vec{v}_1 + d_{52}\vec{v}_2$, so

$c_1\vec{v}_1 + \dots + c_5\vec{v}_5 = k_1\vec{v}_1 + k_2\vec{v}_2$

} these are statements about relations between the columns of rref(A)!

So the same relations hold for the columns $\{\vec{a}_1, \dots, \vec{a}_5\}$ of A! Thus $\{\vec{a}_1, \vec{a}_2\}$ is a basis for CS(A).

Thm:) The columns of A corresponding to pivot columns in RREF(A) form a basis for CS(A).

Ex: $A = \begin{pmatrix} 1 & 2 & -1 & 3 & 0 \\ 1 & 1 & 0 & 4 & 1 \\ 1 & 4 & -3 & 1 & -2 \end{pmatrix}$

$\text{rref}(A) = \begin{pmatrix} 1 & 0 & 1 & 5 & 2 \\ 0 & 1 & -1 & -1 & -1 \\ 0 & 0 & 0 & 0 & 0 \end{pmatrix}$

There are pivots in the first two columns of rref(A),

so $\left\{ \begin{pmatrix} 1 \\ 1 \\ 1 \end{pmatrix}, \begin{pmatrix} 2 \\ 1 \\ 4 \end{pmatrix} \right\}$ is a basis for $CS(A)$.

Alt: Note $CS(A) = RS(A^T)$. So you can also get a basis for $CS(A)$ by row reducing A^T... but, this involves doing a different row reduction.

Thm: $\dim(RS(A)) = \text{rank}(A) = \dim(CS(A))$

Summary of related key ideas

1. Linear combination – sum of scalar multiples.

$$c_1\vec{v}_1 + \ldots + c_n\vec{v}_n$$

2. Vector space – a set where you can "always do" linear combinations.

3. Linear independence – linear combinations are never "unexpectedly zero".

4. Span – set of all linear combinations.

5. Basis – a "nice" set of "just the right number" of vectors...

   a. enough so that they span.

   b. not so many that they can't be independent.

6. Dimension: that "just the right number".

Note, from this discussion, how everything traces back to linear combinations!

Equivalent facts about nonsingular $(n \times n)$ matrices

A is nonsingular

$\Updownarrow$

$\text{rref}(A) = I$

$\Updownarrow$

$\text{rank}(A) = n$

A has existence property

$\Updownarrow$

A has uniqueness property

$\Updownarrow$

A is invertible

$\Updownarrow$

A is a product of elementary matrices

$\Updownarrow$

A represents a row reduction

$\Updownarrow$

$\det(A) \neq 0$

$\Updownarrow$

∥gram det. by cols of A has nonzero "volume"

$\Updownarrow$

columns of A are l.i.

$\Updownarrow$

rows of A are l.i.

$\Updownarrow$

$\dim(CS(A)) = n$

$\Updownarrow$

$\dim(RS(A)) = n$

$\Updownarrow$

$\dim(NS(A)) = 0$

Wronskians

How can you tell if functions (vectors in $F(a,b)$) are l.i. or l.d.?

With polynomials this is easy because we have a finite basis, and can deal with coefficients.

If the functions are not polynomials, this won't work.

Consider $f_1, \dots, f_n \in D^{n-1}(a,b)$. We need to consider possible solutions to

$$c_1 f_1(x) + \dots + c_n f_n(x) = 0$$

(Note, on the right side is the zero function – which has value 0 <u>for all</u> $x \in (a,b)$.)

We consider the following argument. Suppose $f_1, \dots, f_n$ are <u>dependent</u>:

$$\{f_1, \dots, f_n\} \text{ is l.d.} \Rightarrow c_1 f_1 + \dots + c_n f_n = 0 \quad \begin{pmatrix}\text{nontriv.}\\ c_i\end{pmatrix}$$

$$\Rightarrow c_1 f_1' + \dots + c_n f_n' = 0$$

$$\vdots$$

$$\Rightarrow c_1 f_1^{[n-1]} + \dots + c_n f_n^{[n-1]} = 0$$

$$\Rightarrow \begin{pmatrix} f_1(x) & \cdots & f_n(x) \\ f_1'(x) & & f_n'(x) \\ \vdots & & \vdots \\ f_1^{[n-1]}(x) & \cdots & f_n^{[n-1]}(x) \end{pmatrix} \begin{pmatrix} c_1 \\ \vdots \\ c_n \end{pmatrix} = \begin{pmatrix} 0 \\ \vdots \\ 0 \end{pmatrix}$$

has non trivial solutions.

This is a square matrix, so non trivial solutions means that its determinant must be zero.

$$\Rightarrow \det \begin{pmatrix} f_1(x) & \cdots & f_n(x) \\ \vdots & & \vdots \\ f_1^{[n-1]}(x) & \cdots & f_n^{[n-1]}(x) \end{pmatrix} = 0$$

We call this determinant the <u>Wronskian</u> of $\{f_1, \ldots, f_n\}$.

Note that the Wronskian is a <u>function of x</u>.

Thm: $\{f_1, \ldots, f_n\}$ is l.d. $\Rightarrow$ $W(x) = 0$ <u>for all x</u>

Thm: If $W(x) \neq 0$ for <u>any</u> x, then $\{f_1, \ldots, f_n\}$ is l.i.

Ex: Is $\{x, x^2, x^3\}$ l.i.?

$$W(x) = \det\begin{pmatrix} x & x^2 & x^3 \\ 1 & 2x & 3x^2 \\ 0 & 2 & 6x \end{pmatrix} = x(6x^2) - 1(4x^3)$$

$$= 2x^3$$

This happens to be zero for some x, but the point is that this is <u>not</u> the zero <u>function</u>.

So, these functions are <u>l.i.</u>

<u>WARNING</u>: $W(x) = 0 \not\Rightarrow \{f_1, \dots, f_n\}$ l.d.

The argument (l.d. $\Rightarrow W = 0$) appears on first glance to be reversible, but it is <u>not</u>. Specifically:

$$\left(\begin{array}{c} A\vec{c} = \vec{0} \;\; (\text{for all } x) \\ \text{has a nontriv. sol'n } \vec{c} \end{array}\right) \Longrightarrow \left(\det(A) = 0 \;\; (\text{for all } x)\right)$$

This step is <u>not</u> reversible, because $\det(A) = 0$ implies only that for every $\vec{x}$ there is a nontrivial $\vec{c}$ — <u>not</u> that the same $\vec{c}$ will work for each $\vec{x}$.

The book shows a nice counterexample.

In a restricted category though, the result does reverse.

Def: A function $f: \mathbb{R}^1 \to \mathbb{R}^1$ is analytic (or, "real analytic") if its Taylor series converges to itself.

Ex: In single variable calculus courses it is shown that e^x, $\sin x$, $\cos x$ all have Taylor series that converge to themselves. So, these functions are analytic.

Fact: Sums, products, compositions, fractions (with nonzero denoms!) of analytic functions are analytic.

Thm: If $f_1, \ldots, f_n$ are analytic and the Wronskian is identically zero, then $\{f_1, \ldots, f_n\}$ is l.d..

3.1–3.4 Differential Equations, Intro/review

Students should already have seen some differential equations in Math 32 (or equiv.).

Ex:) $\frac{dy}{dx} = f(x)$; $\frac{dy}{dx} = ky$; $\frac{dy}{dx} = \frac{f(x)}{g(y)}$

Note, DE's do not always have solutions. And they do not always have unique solutions even with an initial condition...

Ex:) Consider the initial value problem

$$\frac{dy}{dx} = 3y^{2/3}, \quad y(0) = 0$$

Note that $y(x) = 0$ is a sol'n. But, so is $y(x) = x^3$!

Under some conditions, we can ensure this does not happen.

Thm:) Consider the IVP: $\frac{dy}{dx} = f(x,y) \quad y(x_0) = y_0$

If f <u>and</u> $\frac{\partial f}{\partial y}$ are continuous near (x_0, y_0), then, sufficiently near x_0, the IVP has a unique solution.

Some techniques:

Separable: Some DE's can be put into the form

$$g(y)\,dy = f(x)\,dx$$

Integrate both sides and solve.

Exact: Some DE's can be put into the form

$$\frac{\partial F(x,y)}{\partial x} + \frac{\partial F(x,y)}{\partial y}\frac{dy}{dx} = 0$$

Rewrite as $\frac{d}{dx} F(x,y) = 0$

and integrate to get $F(x,y) = C$

Ex:) $(2x \cos y) - (x^2 \sin y + 2y)\frac{dy}{dx} = 0$

This is exact, with $F(x,y) = x^2 \cos y - y^2$

Integrating Factor: If we have $y' + p(x)y = q(x)$,

multiply both sides by $\mu(x) = e^{\int p(x)\,dx}$

(note that this gives us $\mu'(x) = \mu(x)\,p(x)$

We get

$$\mu y' + \mu p y = \mu q$$

$$\mu y' + \mu' y = \mu q$$

$$(\mu y)' = \mu q$$

$\vdots$

3.6 – Modeling with DE's

Many real situations can be modeled with DE's. These are some of the simplest.

Ex:) An "interest rate" is really a factor multiplied by principal to give the rate that interest is paid.

For example, if you have \$1,000 in an account, with an interest rate of 6%/yr, then at <u>that instant</u>, the rate of change of your balance is (6%/yr)(\$1,000) = \$60/yr. This is an <u>instantaneous rate</u>.

This generalizes to the DE:

$$\frac{dB}{dt} = rB \quad \leftarrow \text{(Natural Growth Equation)}$$

With an initial balance $B(0) = B_0$, the solution is

$$B(t) = B_0 e^{rt}$$

(sometimes this is referred to as "continuous compounding", but it is better to think of this as simply a proper interpretation of rate.)

Ex:) Suppose you have an interest rate r and an additional saving rate of $\$S/yr$. Then we have

$$\frac{dB}{dt} = rB + S$$

Trick: let $z = rB + S$, so $z' = rB'$.

Then $z' = rz$, which is N.G.E.

Ex:) Say we have a (full) 1000 g tank. 5g/min flows in, mixes perfectly, and flows out at the same rate. The input has 3lbs/gal concentration of salt. How does the concentration in the tank change over time?

Let $C(t)$ = concentration, and $Q(t)$ = quantity of salt.

Then
$$\frac{dQ}{dt} = \left(\frac{3\,lbs}{gal}\right)\left(\frac{5\,gal}{min}\right) - \left(C(t)\right)\left(\frac{5\,gal}{min}\right)$$

$$= 15\,\frac{lbs}{min} - \frac{5\,Q}{1000\,gal}\,\frac{gal}{min}$$

Solve with same trick above, for $Q(t)$. Then $C = \frac{Q}{1000}$.

See also Beale's notes on first order DE's.

Higher Order Linear DE's

Def: An nth order linear DE is of the form

$$\underbrace{g_n(x)\, Y^{[n]} + g_{n-1}(x)\, Y^{[n-1]} + \dots + g_1(x)\, Y' + g_0(x)\, Y}_{L(Y)} = g(x)$$

The left side can be viewed as an operator on Y.

It is linear in the sense that

$$L(c_1 Y_1 + c_2 Y_2) = c_1 L(Y_1) + c_2 L(Y_2)$$

(This does not mean that this is a linear fn of x!)

We confirm by direct computation:

$$L(c_1 Y_1 + c_2 Y_2) = g_n (c_1 Y_1 + c_2 Y_2)^{[n]} + \dots + g_0 (c_1 Y_1 + c_2 Y_2)$$

$$= g_n \left(c_1 Y_1^{[n]} + c_2 Y_2^{[n]}\right) + \dots + g_0 (c_1 Y_1 + c_2 Y_2)$$

$$= c_1 \left(g_n Y_1^{[n]} + \dots + g_0 Y_1\right) + c_2 \left(g_n Y_2^{[n]} + \dots + g_0 Y_2\right)$$

$$= c_1 L(Y_1) + c_2 L(Y_2)$$

Some terminology:

- If $g(x) = 0$, we say the equation is <u>homogeneous</u>.
- If we prescribe values for $y, y', \dots, y^{[n-1]}$ at some value x_0, the equation becomes an <u>initial value problem</u>.

Thm: If $q_0, \dots, q_n$ are continuous, g is continuous, and $q_n(x) \neq 0$ for all x on (a,b), then the IVP

$$q_n y^{[n]} + \dots + q_0 y = g(x)$$

$$y(x_0) = k_0, \; \dots, \; y^{[n-1]}(x_0) = k_{n-1}$$

has <u>exactly</u> one solution on (a,b).

(We will refer to this theorem as the <u>"existence and uniqueness theorem"</u> in this course. It is very important, but its proof is beyond the scope of this course.)

Thm: If $g_0, \ldots, g_n$ are continuous and $g_n \neq 0$, then the set of solutions to

$$g_n y^{[n]} + \ldots + g_0 y = 0$$

is an <u>n-dim</u> <u>subspace</u> of <u>C^n</u>.

Partial Pf: Since $g_n \neq 0$ we can write

$$y^{[n]} = \frac{g_{n-1} y^{[n-1]} + \ldots + g_0 y}{g_n}$$

The functions on the right are all continuous, so $y^{[n]}$ is too.

So $y \in C^n$.

To check the solutions form a subspace, say y_1, y_2 are solutions:

1. closed under addition: $L(y_1 + y_2) = L(y_1) + L(y_2) = 0 + 0 = 0$.
2. closed under scalar mult.: $L(cy_1) = cL(y_1) = c \cdot 0 = 0$.

(compare to a previous argument made about sols to $A\vec{x} = \vec{0}$!)

(To show this set is n-dimensional, we will wait until we have a more powerful linear algebra tool called a linear transformation, and related facts.)

A basis for the set of solutions to a hom. lin. DE is called a "<u>fundamental set of solutions</u>".

Ex:) Consider the DE.

$$y'' + y = 0$$

This is hom. and linear, and 2nd order.

So we know that the set of solutions is 2-dim.

Observe that $\sin x$ and $\cos x$ are solutions...

So, <u>every</u> solution is of the form

$$y = c_1 \sin x + c_2 \cos x$$

and $\{\sin x, \cos x\}$ is a fundamental set of sols.

If we have a nonhomogeneous linear DE, we only need to find <u>one</u> particular solution and the homogeneous solutions in order to have all of the solutions.

Thm: Suppose y_p is a solution to the linear DE

$$L(y) = g(x)$$

Suppose that $\{y_1, \ldots, y_n\}$ are a fund. set of sols to the associated hom. linear DE

$$L(y) = 0$$

Then the general solution to the non hom. lin. DE is

$$y = y_p + (c_1 y_1 + \cdots + c_n y_n)$$

Pf:

$$L(y) = L(y_p) + c_1 L(y_1) + \ldots + c_n L(y_n)$$
$$= g(x) + 0 + \ldots + 0$$
$$= g(x).$$

So every function of this form is a solution.

On the other hand, if y is any solution, we have

$$L(y - y_p) = L(y) - L(y_p) = g(x) - g(x) = 0$$

So $y - y_p$ is a solution to the associated hom. lin. DE, and thus

$$y - y_p = c_1 y_1 + \cdots c_n y_n$$

So

$$y = y_p + (c_1 y_1 + \ldots + c_n y_n)$$

Ex:) What is the complete set of solutions to the DE

$$y'' + y = 2e^x$$

First, note that $y_p = e^x$ is a solution.

Second, we already know from prev. example that $\{\sin x, \cos x\}$ are a fund. set of sols to the associated hom. DE

$$y'' + y = 0$$

So, every solution to the given equation is

$$y = e^x + C_1 \sin x + C_2 \cos x$$

Note that there is a very strong similarity between this result about solutions to non hom. DE's and an old result about solutions to non hom. systems of equations!

(Compare these statements, <u>and</u> their proofs!)

Recall that

$$\big(w(x) \neq 0 \text{ for \underline{any} } x\big) \Rightarrow \big(\{f_1, \ldots, f_n\} \text{ is l.i.}\big)$$

But, unfortunately

$$\big(w(x) = 0 \text{ for \underline{all} } x\big) \not\Rightarrow \big(\{f_1, \ldots, f_n\} \text{ is l.d.}\big)$$

It turns out that solutions to hom. lin. DE's are sufficiently nice that this actually works out for these fns.

In fact, we get an even stronger fact.

<u>Thm:</u>) Let $y_1, \ldots, y_n$ be solutions to the nth order lin. hom. DE (satisfying the conditions of the uniqueness/exist. thm.)

$$L(y) = 0$$

Then

$$\big(w(x_0) = 0 \text{ for \underline{any} } x_0\big) \Rightarrow \big(\{y_1, \ldots, y_n\} \text{ is l.d.}\big)$$

<u>Pf:</u>) If $w(x_0) = 0$, then there is a <u>non trivial</u> sol'n $c_1, \ldots, c_n$ to

$$c_1 y_1(x_0) + \cdots + c_n y_n(x_0) = 0$$

$$\vdots$$

$$c_1 y_1^{[n-1]}(x_0) + \cdots + c_n y_n^{[n-1]}(x_0) = 0$$

Consider the function

$$u(x) = c_1 y_1(x) + \cdots + c_n y_n(x)$$

The above equations then tell us that

$$u(x_0)=0, \quad u'(x_0)=0, \quad \ldots, \quad u^{[n-1]}(x_0)=0$$

So u is the solution to the IVP

$$L(y)=0$$

$$y(x_0)=0, \quad y'(x_0)=0, \quad \ldots, \quad y^{[n-1]}(x_0)=0$$

But of course the zero function is the sol. to that IVP.

So we have $\quad u(x)=0$

$$\Rightarrow c_1 y_1(x) + \cdots + c_n y_n(x) = 0$$

So $\{y_1, \ldots, y_n\}$ is l.d.

Recall that for arbitrary functions, the problem was that the c's that work at one value of x are not necessarily the same as the c's that work at another x...

In this proof, it is the uniqueness of sols to IVP's that allows us to get around that problem.

Note the following easy corollary:

Cor: If $\{y_1, \dots, y_n\}$ are solutions to $L(y) = 0$, (nth order, satisfying the conditions of the uniqueness/exist. thm.)

then either

① W is always 0 and the fns are l.d.

or

② W is never 0 and the fns are l.i.

Pf: ② follows from old result about Wronskian.

If ② does not happen then W is 0 for some x_0. Our previous thm tells us then that the fns are l.d., and again our old result about the Wronskian tells us that W is always 0.

Homogeneous Const. Coeff. Linear DE's

In many natural situations, the coefficient functions $q_i(x)$ are physical quantities that are constant.

Ex:) Forced harmonic motion

Mass on a spring in a resistive medium.

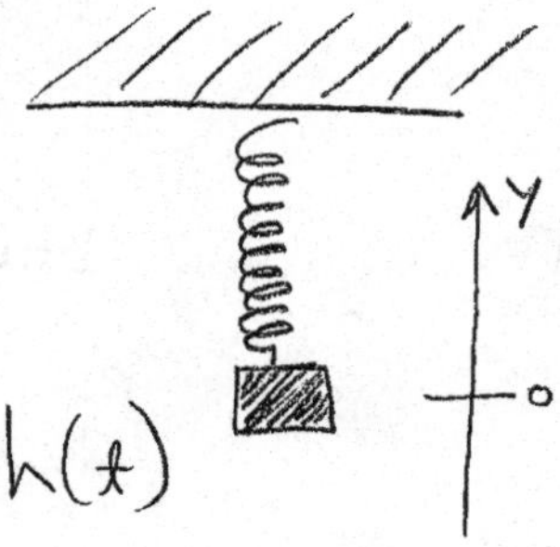

Model: $my'' + \mu y' + ky = h(t)$

(m: mass, μ: friction coefficient, k: spring constant)

Ex:) LRC circuits

A circuit including an inductor, a resistor, and a capacitor.

Model: $LI'' + RI' + \frac{1}{C} I = E'$

(L: inductance, R: resistance, C: capacitance)

Our equation has the form

$$a_n y^{[n]} + \cdots + a_0 y = 0$$

Consider solutions of the form

$$y(x) = e^{\lambda x}$$

The equation becomes

$$\left(a_n \lambda^n + \cdots + a_0\right) e^{\lambda x} = 0$$

This gives us solutions <u>iff</u>

$$a_n \lambda^n + \cdots + a_0 = 0$$

This is called the <u>characteristic polynomial</u>. We consider three cases of types of roots – distinct real, repeated real, complex.

But we must know how to find roots of polynomials!

Finding Roots

① Quadratic equation

② Factor Theorem: a is a root of polynomial $p(x)$

iff $(x-a)$ is a factor of polynomial $p(x)$.

(see proof on class webpage)

We often find roots by factoring!

③ Rational root theorem

If $f(x) = a_n x^n + \cdots + a_1 x + a_0$ has integer coeffs

and if $r = p/q$ is a rational root, then

$p \mid a_0$ and $q \mid a_n$

④ Polynomial division — If $p(x) = (x-a)\,q(x)$, then

you can look for factors/roots of q instead of p...

⑤ Intermediate value theorem — If $p(a)$ and $p(b)$

have opposite signs, there must be a root between

a and b.

Ex: What is the quotient $q(x)$ below?

$$x^4 - 4x^3 + 9x^2 - 7x - 6 = (x-2)\,q(x)$$

$$
\begin{array}{r l}
 & x^3 - 2x^2 + 5x + 3 \\
x-2 \,\big)\!\! & \overline{x^4 - 4x^3 + 9x^2 - 7x - 6} \\
 & \underline{x^4 - 2x^3} \\
 & \quad -2x^3 + 9x^2 - 7x - 6 \\
 & \quad \underline{-2x^3 + 4x^2} \\
 & \qquad\quad 5x^2 - 7x - 6 \\
 & \qquad\quad \underline{5x^2 - 10x} \\
 & \qquad\qquad\quad 3x - 6 \\
 & \qquad\qquad\quad \underline{3x - 6} \\
 & \qquad\qquad\qquad\; 0
\end{array}
$$

NB – "synthetic division" is <u>not allowed in this course</u>. (Students usually do not understand why it works! By comparison, polynomial division is relatively transparent.)

Ex:) Solve the D.E.

$$y'' - 3y' + 2y = 0$$

The char. poly. is

$$\lambda^2 - 3\lambda + 2 = 0$$

$$\Rightarrow (\lambda - 2)(\lambda - 1) = 0$$

$$\Rightarrow \lambda = 1, 2$$

So e^x, e^{2x} are solutions.

Also – as these are indep. and we know there is a 2-dim space of solutions, we have that $\{e^x, e^{2x}\}$ is a fund. set of sols.

Ex:) Solve

$$y'''' + 8y''' + 3y'' - 32y' - 28y = 0$$

Char. Poly. is

$$p(\lambda) = \lambda^4 + 8\lambda^3 + 3\lambda^2 - 32\lambda - 28 = 0$$

-1 is a root! And

$$p(\lambda) = (\lambda + 1)(\lambda^3 + 7\lambda^2 - 4\lambda - 28)$$

So now we can look for roots of

$$g_1(\lambda) = \lambda^3 + 7\lambda^2 - 4\lambda - 28$$

Note that 2 is a root, and

$$q_1(\lambda) = (\lambda-2)(\lambda^2+9\lambda+14)$$

Then $q_2(\lambda) = \lambda^2+9\lambda+14$ is easy to factor,

$$q_2(\lambda) = (\lambda+7)(\lambda+2)$$

So we have

$$p(\lambda) = (\lambda+1)(\lambda-2)(\lambda+7)(\lambda+2)$$

and then solutions $e^{-x}, e^{2x}, e^{-7x}, e^{-2x}$

And $\{e^{-x}, e^{2x}, e^{-7x}, e^{-2x}\}$ is a fund. set of sols because they are 4 fns that are l.i., in a 4-dim. set of sols.

Distinct Real Roots

If roots are distinct, then:

$$\begin{aligned}\text{\# of solutions } e^{\lambda x} &= \text{\# of roots}\\ &= \text{degree of char. poly}\\ &= \text{order of CCLDE}\\ &= \text{dim of space of sols}\\ &= \text{\# of fns in a fund sol. set}\end{aligned}$$

These functions are l.i., so they are a fund. sol. set.

Real Repeated Roots

If a root is repeated, then we don't get enough solutions...

Ex:) Solve the CCLDE:

$$y'' - 2y' + y = 0$$

The char. poly. is $\lambda^2 - 2\lambda + 1 = 0$

$$(\lambda - 1)^2 = 0 \Rightarrow \lambda = 1$$

So we know e^{1x} is a solution; but we also know that we still need to find one more solution...

One might suspect that since 1 is a root "twice", the other solution *might* be related to e^{1x}

We *try* a function of the form

$$y = u(x)\,e^{1x}$$

Plugging this in gives us

$$(ue^x)'' - 2(ue^x)' + (ue^x) = 0$$

$$\vdots$$

$$u''e^x = 0$$

$$\Rightarrow u'' = 0$$

Conveniently $u(x) = x$ satisfies this. So

$$y = xe^x$$

is another solution.

Observe, 1 was repeated once (multiplicity = 2), and we needed (and found) one new solution.

This good news generalizes.

Def: If the complete factorization of $p(x)$ is

$$p(x) = (x-r_1)^{m_1} \cdots (x-r_k)^{m_k}$$

then we say that each root r_i has <u>multiplicity</u> m_i.

Thm: Let $p(\lambda)$ be the char. poly. of a CCLDE, and r a root of multiplicity m. Then

$$e^{rx}, xe^{rx}, \ldots, x^{m-1}e^{rx}$$

are solutions to the homogeneous CCLDE.

We then get the m solutions that we "should" have for this root.

Will these solutions be independent?

Ex:) Consider $\{e^{-x}, e^{2x}, xe^{2x}, e^{3x}, xe^{3x}, x^2e^{3x}\}$.

Suppose $c_1e^{-x} + c_2e^{2x} + c_3xe^{2x} + c_4e^{3x} + c_5xe^{3x} + c_6x^2e^{3x} = 0$

Multiplying through by e^{-3x} and taking $\lim_{x\to\infty}$, we get

$$\lim_{x\to\infty}\left(\underbrace{c_1e^{-4x} + c_2e^{-2x} + c_3xe^{-2x}}_{\text{these limits are zero}} + c_4 + c_5x + c_6x^2\right) = 0$$

$$\lim_{x\to\infty}\left(c_4 + c_5x + c_6x^2\right) = 0$$

$$\Rightarrow c_4, c_5, c_6 = 0$$

So our original linear combination becomes

$$c_1e^{-x} + c_2e^{2x} + c_3xe^{2x} = 0$$

Now multiplying through by e^{-2x} and taking $\lim_{x\to\infty}$, we get

$$\lim_{x\to\infty}\left(c_1e^{-3x} + c_2 + c_3x\right) = 0$$

(this limit is zero)

$$\lim_{x\to\infty}\left(c_2 + c_3x\right) = 0$$

$$\Rightarrow c_2, c_3 = 0$$

This leaves us with just

$$c_1e^{-x} = 0$$

$$\Rightarrow c_1 = 0$$

So $c_1, \ldots, c_6$ are all zero, and thus the functions are l.i..

This method always works for these sorts of functions!

Thm:) Consider the CCLDE $L(y)=0$, with char. poly. $p(\lambda)$.

If $p(\lambda)$ has all real roots, giving solutions of the forms

$$e^{r_i x} \quad \text{and} \quad x^k e^{r_j x}$$

then these solutions are linearly independent.

Pf:) Suppose $c_1 y_1 + \cdots + c_m y_m = 0$. Let r_1 be the greatest root of $p(\lambda)$, and consider the fact that

$$\lim_{x\to\infty} e^{-r_1 x}\left(c_1 y_1 + \ldots + c_m y_m\right) = 0$$

On the LHS, solutions corresponding to roots other than r_1 will give terms that have $\lim = 0$ because of the exponential. Removing those terms yields

$$\lim_{x\to\infty} e^{-r_1 x}\left(k_0 e^{r_1 x} + k_1 x e^{r_1 x} + \cdots + k_i x^i e^{r_1 x}\right) = 0$$

$$\lim_{x\to\infty} \left(k_0 + k_1 x + \ldots + k_i x^i\right) = 0$$

So all of these coefficients must be 0.

Repeating this process for each root (in decreasing order), we conclude that all $c_1, \ldots, c_m$ must be 0. So the solutions are independent.

■

Complex numbers

Recall, $i = \sqrt{-1}$. Multiples of i are called "imaginary".

Numbers of the form $a+bi$ (where $a,b \in \mathbb{R}$) are "complex".

Complex numbers add and multiply in the expected ways:

$$(a+bi)+(c+di) = (a+c)+(b+d)i$$

$$(a+bi)(c+di) = (ac-bd)+(ad+bc)i$$

A surprising result of Taylor series is:

$$e^{i\theta} = (\cos\theta) + (\sin\theta)\,i$$

The angle interpretation of θ and the trig in the above equation suggest thinking of an imaginary axis $\perp$ to the real axis:

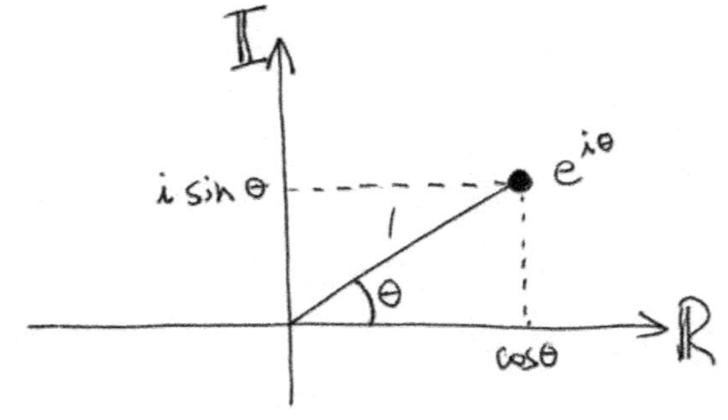

This is called the "complex plane". Note, every complex number can be written in the form $re^{i\theta}$.

Many algebraic facts about complex numbers and functions have appealing geometric interpretations on the complex plane.

For example, complex multiplication:

$$\left(r_1 e^{i\theta_1}\right)\left(r_2 e^{i\theta_2}\right) = \left(r_1 r_2\right) e^{i(\theta_1+\theta_2)}$$

Geometrically, we note that lengths multiply and angles add.

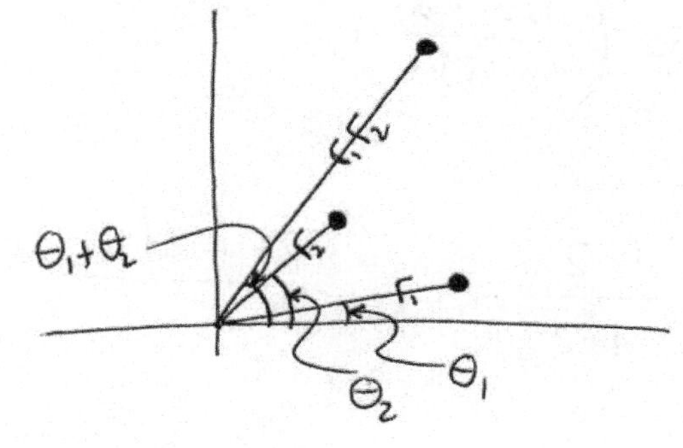

The "complex conjugate" of a complex number $(a+bi)$ is the number $(a-bi)$. It is the reflection through the real axis.

$z = a+bi$

$\bar{z} = a-bi$

Note: ① $z = re^{i\theta} \Rightarrow \bar{z} = re^{-i\theta}$ ③ $\overline{z_1+z_2} = \bar{z}_1 + \bar{z}_2$

② $(\bar{z})^k = \overline{(z^k)}$ ④ $\overline{z_1 z_2} = \bar{z}_1 \bar{z}_2$

Complex conjugates are very useful in dividing complex numbers in rectangular form.

$$\frac{a+bi}{c+di} = \frac{a+bi}{c+di}\left(\frac{c-di}{c-di}\right) = \frac{(ac+bd)+(bc-ad)i}{\underbrace{c^2+d^2}}$$

↖ (real!)

$$= \left(\frac{ac+bd}{c^2+d^2}\right) + \left(\frac{bc-ad}{c^2+d^2}\right)i$$

Ex: $\frac{3+2i}{1+2i} = \frac{3+2i}{1+2i}\,\frac{1-2i}{1-2i} = \frac{7-4i}{5}$

Some useful facts about polynomials:

① Every polynomial with complex coefficients will factor <u>completely</u> (product of linear factors)

Ex:) $x^2+1 = (x-i)(x+i)$

Ex:) $x^3-x^2+2 = (x+1)(x-(1+i))(x-(1-i))$

② If $p(x)$ has <u>real</u> coefficients and if z is a root, then $\bar{z}$ is a root also.

(see above examples)

Pf:) $p(\bar{z}) = \sum a_k (\bar{z})^k = \sum a_k \overline{(z^k)} = \sum \bar{a}_k \overline{(z^k)}$

$= \sum \overline{a_k z^k} = \overline{\sum a_k z^k} = \overline{p(z)} = \bar{0} = 0.$

Another useful fact:

If $a+bi = c+di$, and a,b,c,d are all real, then $a=c$ and $b=d$.

(Equiv.: 1 and i are l.i.)

Note, we <u>need</u> a,b,c,d to be <u>real</u> for this to work...

Another useful fact: Every nonzero complex number has exactly n nth roots.

Roots of unity

Let $\rho = e^{2\pi i/n}$. Then $1, \rho, \rho^2, \ldots, \rho^{n-1}$ are nth roots of 1.

$$(\rho^k)^n = (\rho^n)^k = (e^{2\pi i})^k = 1^k = 1$$

Finding the nth root of α

Let r be any such root.

(E.g., if $\alpha = me^{i\theta}$, you could use $r = (m^{1/n})e^{i\theta/n}$)

Then the n nth roots of α are: $r, r\rho, r\rho^2, \ldots, r\rho^{n-1}$.

$$(r\rho^k)^n = r^n(\rho^k)^n = \alpha \cdot 1 = \alpha$$

Ex: Find the three cube roots of i.

$\alpha = i = 1e^{i\pi/2}$, so we can use $r = 1e^{i\pi/6}$, $\rho = e^{2\pi i/3}$.

Then the three cube roots are:

$$r = 1e^{\pi i/6} = e^{i(\pi/6)} = \frac{\sqrt{3}}{2} + i\left(\frac{1}{2}\right)$$

$$r\rho = 1e^{\pi i/6}e^{2\pi i/3} = e^{i(5\pi/6)} = \frac{-\sqrt{3}}{2} + i\left(\frac{1}{2}\right)$$

$$r\rho^2 = 1e^{\pi i/6}e^{4\pi i/3} = e^{i(3\pi/2)} = 0 + i(-1)$$

Complex valued solutions

Thm: If $L(y) = 0$ is a CCLDE with real coefficients and if $y(x) = u(x) + i v(x)$ is a complex valued solution, ($u(x)$, $v(x)$ real)

then $u(x)$ and $v(x)$ are also solutions.

Pf: $0 = L(y) = L(u + iv)$ ← by linearity

$= L(u) + i L(v)$

these are real, because u, v are real and L has real coefficients.

$= 0 + i \, 0$

So $L(u) = 0$, $L(v) = 0$ and thus u, v are themselves solutions ■

So if you find <u>one</u> complex valued solution, it gives you <u>two</u> real valued solutions also!

Applying this observation to the solution (with $\lambda = r = a+bi$)

$$y = e^{rx} = e^{(a+bi)x} = e^{ax} e^{bix}$$
$$= \left(e^{ax} \cos(bx)\right) + i\left(e^{ax} \sin(bx)\right)$$

we can conclude two real solutions,

$$e^{ax} \cos(bx) \quad \text{and} \quad e^{ax} \sin(bx)$$

Does this give us "too many" solutions? No — because if $r = a+bi$ is a root, remember that the conjugate $\bar{r} = a - bi$ is a root also, and gives us the <u>same two</u> real solutions.

<u>Thm:</u> If $L(y) = 0$ is a real CCLDE and $r, \bar{r}$ are a pair of roots of the char. poly., then

$$e^{ax} \cos(bx) \quad \text{and} \quad e^{ax} \sin(bx)$$

are real solutions, independent, and with the same span as the solutions

$$e^{rx} \quad \text{and} \quad e^{\bar{r}x}$$

If the characteristic polynomial has some complex roots, will the set of solutions we will find by these methods be independent? (And thus a fundamental set, because we already know we will have the right number...) Yes!

Thm: If $L(y)=0$ has characteristic polynomial $p(\lambda)$ with roots $r_1, \ldots, r_n$, and we form a set of solutions by:

① For real roots r of multiplicity m we include

$$e^{rx}, xe^{rx}, \ldots, x^{m-1}e^{rx}$$

② For complex conjugate pairs of roots $r=a+bi$, $\bar{r}=a-bi$ of multiplicity m we include

$$e^{ax}\cos bx, e^{ax}\sin bx, \ldots, x^{m-1}e^{ax}\cos bx, x^{m-1}e^{ax}\sin bx$$

then this set of functions is independent, and is a fundamental set of solutions.

We will not prove this theorem in this class.

Ex:) Solve the CCLDE:

$$y''' - y'' + 2y = 0$$

The char. poly is

$$p(\lambda) = \lambda^3 - \lambda^2 + 2$$
$$= (\lambda+1)(\lambda-(1+i))(\lambda-(1-i))$$

with roots -1, $1+i$, $1-i$

This gives us independent solutions

$$\{e^{-x}, e^x \cos x, e^x \sin x\}$$

Ex:) Solve the CCLDE:

$$y'''' - 4y''' + 8y'' - 8y' + 4y = 0$$

The char. poly. is

$$p(\lambda) = \lambda^4 - 4\lambda^3 + 8\lambda^2 - 8\lambda + 4$$
$$= (\lambda - (1+i))^2 (\lambda - (1-i))^2$$

← (NB, this is hard to factor...)

We have roots $r = (1+i)$ and $\bar{r} = (1-i)$, each with multiplicity 2. So solutions are

$$\{e^{rx}, e^{\bar{r}x}, xe^{rx}, xe^{\bar{r}x}\}$$

or

$$\{e^x \cos x, e^x \sin x, xe^x \cos x, xe^x \sin x\}$$

Ex:) Solve the IVP :

$$y''' - y'' + 2y = 0 \quad , \quad \begin{array}{l} y(3) = 1 \\ y'(3) = 5 \\ y''(3) = 8 \end{array}$$

We already know the general solution from a previous example, so we need only solve for the constants.

We have

$$y = c_1 e^{-x} + c_2 e^{x} \cos x + c_3 e^{x} \sin x$$

$$y' = -c_1 e^{-x} + c_2 e^{x} \cos x - c_2 e^{x} \sin x + c_3 e^{x} \sin x + c_3 e^{x} \cos x$$

$$= (-c_1) e^{-x} + (c_2 + c_3) e^{x} \cos x + (c_3 - c_2) e^{x} \sin x$$

$$y'' = c_1 e^{-x} + (2c_3) e^{x} \cos x + (-2c_2) e^{x} \sin x$$

So at $x = 3$, we have

$$1 = (e^{-3}) c_1 + (e^{3} \cos(3)) c_2 + (e^{3} \sin(3)) c_3$$

$$5 = (-e^{-3}) c_1 + (e^{3}(\cos(3) - \sin(3))) c_2 + (e^{3}(\cos(3) + \sin(3))) c_3$$

$$8 = (e^{3}) c_1 + (-2e^{3} \sin(3)) c_2 + (2e^{3} \cos(3)) c_3$$

We could solve this system for c_1, c_2, c_3, but note that the coefficients are very inconvenient.

Instead we substitute $t = x-3$ and rewrite as

$$y''' - y'' + 2y = 0$$

(derivs w.r.t. t; note $\frac{dy}{dx} = \frac{dy}{dt}$)

$y(0) = 1$

$y'(0) = 5$

$y''(0) = 8$

values of t, when $x=3$

The general solution has the same form, as do derivs:

$$y = d_1 e^{-t} + d_2 e^{t}\cos t + d_3 e^{t}\sin t$$

$$y' = (-d_1)e^{-t} + (d_2+d_3)e^{t}\cos t + (d_3-d_2)e^{t}\sin t$$

$$y'' = d_1 e^{-t} + (2d_3)e^{t}\cos t + (-2d_2)e^{t}\sin t$$

At $x=3$, which is $t=0$, we can solve for d_1, d_2, d_3 with

$$1 = (1)d_1 + (1)d_2 + (0)d_3$$

$$5 = (-1)d_1 + (1)d_2 + (1)d_3$$

$$8 = (1)d_1 + (0)d_2 + (2)d_3$$

You can then back substitute $t = x-3$.

NB, this solution is then written in terms of a <u>different basis</u> than on the previous page.

That's okay!

4.3 – Undetermined Coefficients

For nonhomogeneous equations, recall that we need only to find <u>any single particular solution</u>; we can then get the complete solution by adding the general solution to the assoc. hom. equation.

"Undetermined coefficients" is basically a guess-and-check method, advised by experience.

Ex i) Find a particular solution to

$$y'' + 2y' - 3y = 4e^{2x}$$

Note that if we chose $y_p = Ae^{2x}$, then y' and y'' would have similar forms, so the entire LHS would be a multiple of e^{2x}, as needed. We can then solve for (determine) the as yet "undetermined coefficient" A.

$$y_p = Ae^{2x}$$

$$(4Ae^{2x}) + 2(2Ae^{2x}) - 3(Ae^{2x}) = 4e^{2x}$$

$$5Ae^{2x} = 4e^{2x}$$

$$A = \frac{4}{5}$$

$$y_p = \frac{4}{5}e^{2x}$$

Ex: Find a particular solution to

$$y'' + 2y' - 3y = x^2$$

We would have a shot at balancing this equation with

$$y = Ax^2$$

but, ... y' and y'' then take forms not on RHS... ☹

But if we put those forms in y too, then that could fix it:

$$y_p = Ax^2 + Bx + C$$

(Note, the derivs of these new terms are of the same forms, so no new terms are suggested.)

Plugging this in to the DE, we get

$$(2A) + 2(2Ax + B) - 3(Ax^2 + Bx + C) = x^2$$

$$(-3A)x^2 + (4A - 3B)x + (2A + 2B - 3C) = x^2$$

$$A = -1/3 \qquad B = -4/9 \qquad C = -14/27$$

So a particular solution is

$$y_p = -\frac{1}{3}x^2 - \frac{4}{9}x - \frac{14}{27}$$

Ex:) What form would be natural to try for a particular solution to

$$y'' + 2y' - 3y = 3\sin x$$

It appears we will need $\sin x$ on the LHS; but that will also create $\cos x$ on the LHS. But if we use both $\sin x$ and $\cos x$, we can hope to balance...

$$y_p = A\cos x + B\sin x$$

This form works, with $A = \frac{-3}{10}$, $B = \frac{-3}{5}$

Ex:) What form should we try for

$$y'' + 2y' - 3y = -2e^x \cos x$$

Well, using $Ae^x\cos x$ creates a need for $Be^x\sin x$; this term does not suggest any others. You can check that

$$y_p = Ae^x\cos x + Be^x\sin x$$

works, with $A = 2/17$, $B = -8/17$

Ex: Find a particular solution for

$$y'' + 2y' - 3y = -2e^x$$

Tempting to try Ae^x... but, note Ae^x is a solution to the homogeneous equation, for all A!

Remember that xe^x has derivative terms with e^x...

So we might try $y_p = Axe^x$:

$$\left(A(x+2)e^x\right) + 2\left(A(x+1)e^x\right) - 3\left(Axe^x\right) = -2e^x$$

$$4Ae^x = -2e^x$$

$$A = -1/2$$

$$y_p = -\frac{1}{2}xe^x$$

Moral: If you know the homogeneous solutions, you know what forms not to bother trying.

The following theorem makes recommendations of forms to try:

Thm: Consider the CCLDE

$$L(y) = a_n y^{[n]} + \cdots + a_1 y' + a_0 y = g(x)$$

with char. poly. $p(\lambda)$.

If $g(x) = Ax^k e^{ax}\cos bx + Bx^k e^{ax}\sin bx \quad (k \in \mathbb{N})$

then ① If $r = a+bi$ is not a root of $p(\lambda)$, try

$$y_p = (c_k x^k + \cdots + c_0)e^{ax}\cos(bx) + (d_k x^k + \cdots + d_0)e^{ax}\sin(bx)$$

② If $r = a+bi$ is a root of $p(\lambda)$ of mult. m, try

$$y_p = x^m(c_k x^k + \cdots + c_0)e^{ax}\cos(bx) + x^m(d_k x^k + \cdots + d_0)e^{ax}\sin(bx)$$

Note ① $r = a+bi = 0$ gives $g(x) = Ax^k$

② $b = 0$ gives $g(x) = Ax^k e^{ax}$ with $r = a$

③ $a = 0$ gives $g(x) = Ax^k\cos(bx) + Bx^k\sin(bx)$ with $r = bi$

Ex i) What form should we try for a particular solution to

$$y'' - 2y' + 5y = x e^x \cos(2x)$$

The char. poly is

$$p(\lambda) = \lambda^2 - 2\lambda + 5$$
$$= \big(\lambda - (1+2i)\big)\big(\lambda - (1-2i)\big)$$

Our $g(x)$ is of the form

$$x e^x \cos(2x) = x^k e^{ax} \cos(bx)$$

with $k=1$, $r = a+bi = 1+2i$, and r is a root of $p(\lambda)$ with $m=1$.

So we try

$$y_p = x\left(c_1 x + c_0\right) e^x \cos(2x)$$
$$+ x\left(d_1 x + d_0\right) e^x \sin(2x)$$

$$= c_1 x^2 e^x \cos(2x) + c_0 x e^x \cos(2x)$$
$$+ d_1 x^2 e^x \sin(2x) + d_0 x e^x \sin(2x)$$

If there are multiple terms on the RHS that do not fit the same form, you can do "one at a time".

That is, if $L(y_{p_1}) = g_1$

$L(y_{p_2}) = g_2$

then $L(y_{p_1} + y_{p_2}) = g_1 + g_2$

Ex:) Find a particular solution to

$$L(y) = y'' + 2y' + y = x^2 + e^x$$

The equation $L(y) = x^2$ has a particular solution

$$y_{p_1} = Ax^2 + Bx + C$$

The equation $L(y) = e^x$ has a particular solution

$$y_{p_2} = De^x$$

So, the particular solution we are looking for is of the form

$$y_p = Ax^2 + Bx + C + De^x$$

4.5 - Applications

Mass on a spring

Say a mass is attached to a fixed object by a spring, subject to forces from the spring, friction/damping, and external forces. What is the position u as a function of time t?

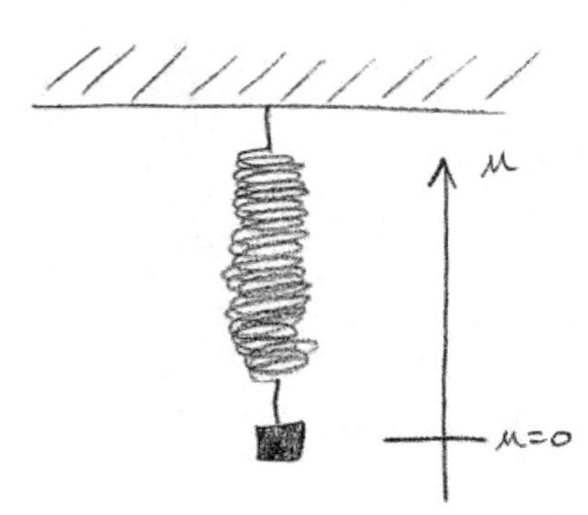

Forces: Hooke's law $F = -ku$

friction $F = -fu'$

external $F = h(t)$

Total forces:

$$F = -ku - fu' + h(t)$$

Recall that $F = ma = mu''$. So this becomes

$$mu'' = -ku - fu' + h(t)$$

$$mu'' + fu' + ku = h(t)$$

As previously claimed, note that this is a CCLDE.

Unforced cases ($h=0$)

① No friction ($f=0$; "undamped")

$$mu'' + ku = 0$$

Characteristic polynomial is $p(\lambda) = m\lambda^2 + k$

$$\Rightarrow \lambda = \pm i\sqrt{\tfrac{k}{m}} = \pm i\omega_0 \quad \left(\omega_0 = \sqrt{k/m}\right)$$

So the f.s.s. is $\{\cos\omega_0 t, \sin\omega_0 t\}$

and gen. sol. is $u = C_1\cos\omega_0 t + C_2\sin\omega_0 t$

Recalling high school trig, you can rewrite as

$$u = A\cos(\omega_0 t - \phi)$$

$$= (A\cos\phi)\cos\omega_0 t + (A\sin\phi)\sin\omega_0 t$$

You can find A, ϕ by interpreting $C_1 = A\cos\phi$, $C_2 = A\sin\phi$ as

so $A^2 = C_1^2 + C_2^2$

and $\cos\phi = C_1/A$

$\sin\phi = C_2/A$

This is called the "natural frequency"; recall of course that strictly speaking the "frequency" is $\omega_0/2\pi$.

② Friction (damped)

$$m u'' + f u' + k u = 0$$

Char. poly. is

$$m\lambda^2 + f\lambda + k = 0$$

$$\Rightarrow \lambda = \frac{-f \pm \sqrt{f^2 - 4mk}}{2m}$$

Observe that : – this simplifies to previous formula when $f = 0$

– the $\mathrm{Re}(\lambda)$ is <u>always</u> < 0

ⓐ Under damped

If $f^2 - 4mk < 0$....

then we have two complex roots

$$\lambda = -a \pm b i \qquad \left(\begin{array}{l} a = f/2m \\ b = \dfrac{\sqrt{f^2 - 4mk}}{2m} \end{array} \right)$$

and then solutions

$$u = C_1 e^{-at} \cos bt + C_2 e^{-at} \sin bt$$

or

$$u = e^{-at} \left(A \cos(bt - \phi) \right)$$

This is a decaying oscillation.

(b) <u>Over-damped</u>

If $f^2 - 4mk > 0$ ....

then we have two real roots $r_1, r_2 < 0$

and solutions

$$u = c_1 e^{r_1 t} + c_2 e^{r_2 t}$$

This is decaying with no oscillations

(c) <u>Critically damped</u>

If $f^2 - 4mk = 0$ ....

then we have one real root $r = \frac{-f}{2m}$

and solutions

$$u = c_1 e^{rt} + c_2 t e^{rt}$$

I
undamped
R

I
under damped
R

I
critically damped
R

I
overdamped
R

Forced cases

① No friction (f=0; undamped)

$$m u'' + k u = h(t)$$

We consider a case where h is sinusoidal, $\omega \neq \omega_0$.

$$u'' + \omega_0^2 u = a \cos \omega t$$

(recall $\omega_0 = \sqrt{k/m}$)

("forcing frequency"; freq $= \omega / 2\pi$)

We can get all of the solutions by:

homogeneous sols : $u_H = c_1 \cos \omega_0 t + c_2 \sin \omega_0 t$

particular sol. : $u_p = \frac{a}{\omega_0^2 - \omega^2} \cos \omega t$

and $u = u_H + u_p$. (Do these for yourself!)

To see what these solutions look like, let's pick one by setting initial conditions $u(0)=0$, $u'(0)=0$, which gives

$$u = \frac{a}{\omega_0^2 - \omega^2} \left(\cos \omega t - \cos \omega_0 t \right)$$

A useful idea in trig:

$$\cos(a-b) = \cos a \cos b + \sin a \sin b$$

$$\cos(a+b) = \cos a \cos b - \sin a \sin b$$

Subtracting, we get

$$\cos(a-b) - \cos(a+b) = 2 \sin a \sin b$$

In this case then, a difference of two sinusoids (same amplitude, different freqs) is a product of sinusoids.

(Can get similar identities by adding; or using angle addition formulas for sin instead of cos.)

To understand $\cos \omega t - \cos \omega_0 t$, then, we set

$$\begin{aligned} a-b &= \omega t \\ a+b &= \omega_0 t \end{aligned} \quad \Rightarrow \quad a = \frac{\omega_0 + \omega}{2} t \,, \quad \frac{\omega_0 - \omega}{2} t$$

The above identity then gives us

$$\cos \omega t - \cos \omega_0 t = 2 \sin\left(\frac{\omega_0 - \omega}{2} t\right) \sin\left(\frac{\omega_0 + \omega}{2} t\right)$$

If $\omega \approx \omega_0$, then this is a product of

$2 \sin\left(\frac{\omega_0 - \omega}{2} t\right)$ = a slowly changing "amplitude"

$\sin\left(\frac{\omega_0 + \omega}{2} t\right)$ = a fast oscillation.

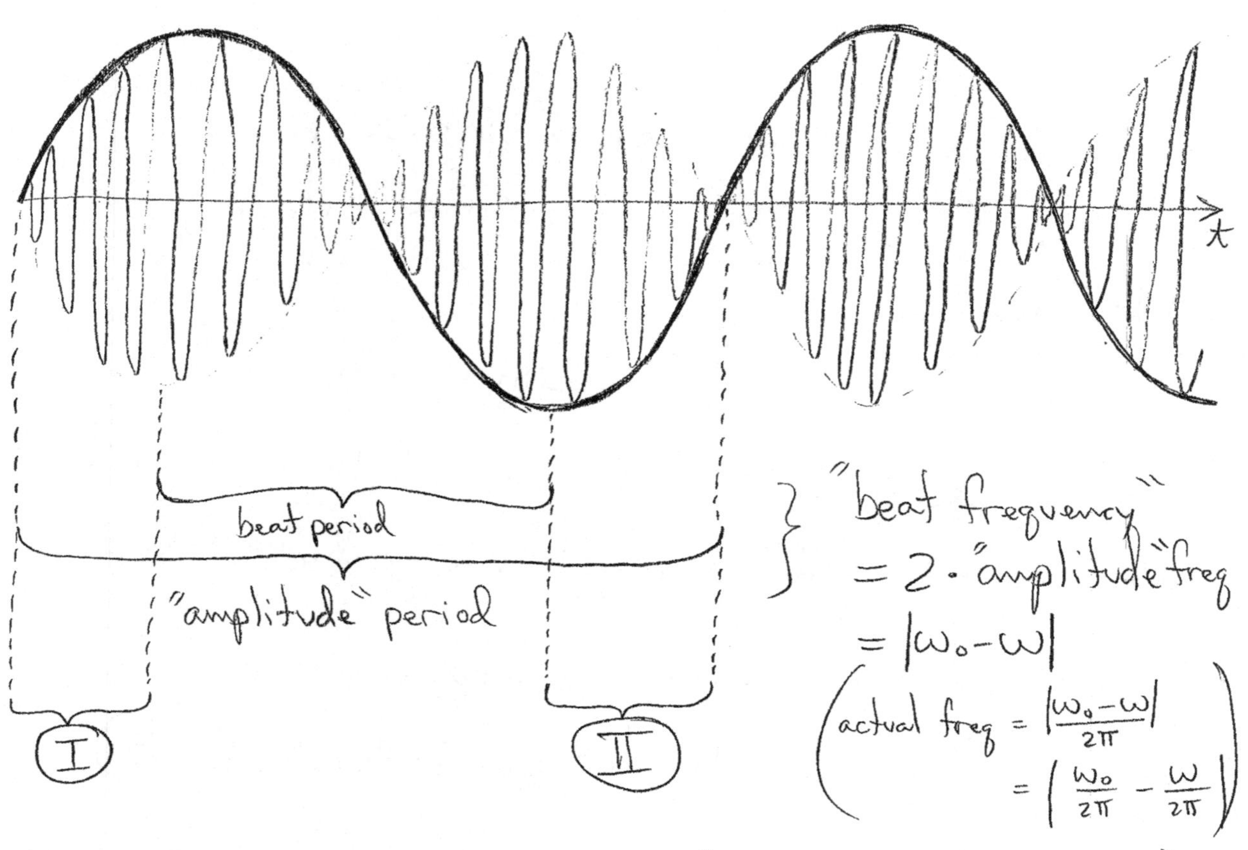

Physical intuition: frequencies (oscillation and forcing) go in and out of phase.

- When they are in phase, amplitude increases (I above).
- When they are out of phase, amplitude decreases (II above).

(An example of a beat frequency involves playing two close notes on a piano; the "rattle" is the beat.)

But what if $\omega_0 = \omega$?

Previous particular solution fails; undetermined coefficients (note $a+bi = \omega i$ <u>is</u> a root of $p(\lambda)$!) results in particular solution below (check!).

$$u_p = \left(\frac{a}{2\omega_0} t\right) \sin \omega_0 t$$

The amplitude grows with time t!

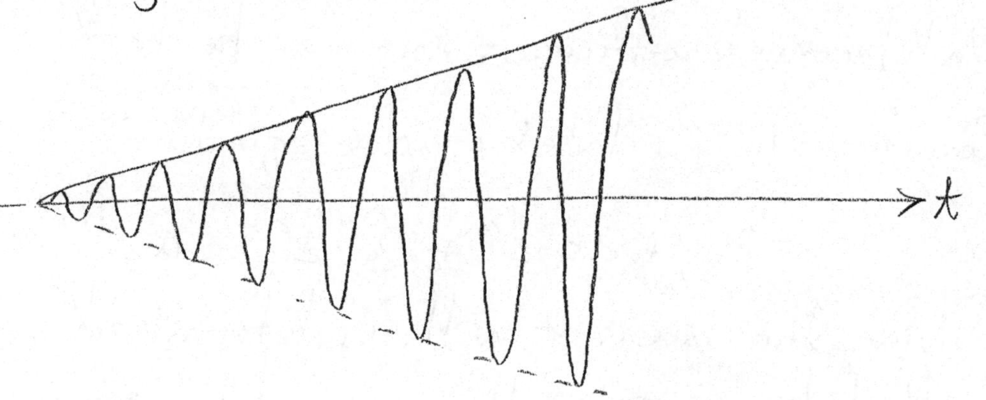

(Forcing and oscillating frequencies never go out of phase!)

This is called resonance. It is very dangerous!

(See London Millenium Bridge (but <u>not</u> Tacoma Narrows!))

② Friction (damped)

We already know the homogeneous solutions; they all decay exponentially and thus do not affect long-term behavior. They are called "transients".

Let's look at particular solutions. Consider

$$u'' + 2cu' + \omega_0^2 u = a\cos\omega t$$

Undetermined coefficients here can be hard...

Consider the "related complex equation"

$$z'' + 2cz' + \omega_0^2 z = ae^{i\omega t}$$

Solving this is easier with undetermined coefficients, and helps us solve the original equation.

Suppose $z = u + iv$; recall $ae^{i\omega t} = a\cos\omega t + i\,a\sin\omega t$

Then $$L(z) = ae^{i\omega t}$$

$$L(u) + iL(v) = a\cos\omega t + i\,a\sin\omega t$$

and so $$L(u) = a\cos\omega t$$
$$L(v) = a\sin\omega t$$

So $u = \mathrm{Re}(z)$ solves the original equation!

Alt:

Suppose $z(t) = u(t) + iv(t)$ is a solution to

$$z'' + 2cz' + \omega_0^2 z = ae^{i\omega t}$$

Expanding by real & imaginary components, we get

$$\begin{pmatrix} u \\ + \\ iv \end{pmatrix}'' + 2c \begin{pmatrix} u \\ + \\ iv \end{pmatrix}' + \omega_0^2 \begin{pmatrix} u \\ + \\ iv \end{pmatrix} = \begin{pmatrix} a\cos\omega t \\ + \\ i\,a\sin\omega t \end{pmatrix}$$

Recall, $\left.\begin{matrix} a+bi = c+di \\ a,b,c,d \text{ real} \end{matrix}\right\} \Longrightarrow \begin{cases} a=c \\ b=d \end{cases}$

Then

$$\begin{cases} u'' + 2cu' + \omega_0^2 u = a\cos\omega t \\ v'' + 2cv' + \omega_0^2 v = a\sin\omega t \end{cases}$$

The first equation above confirms that $u = Re(z)$ is the solution we are looking for.

(Note, similarly, if RHS were $a\sin\omega t$, we could use the same complex equation, and select $v = Im(z)$ as our solution.)

So let's solve

$$z'' + 2cz' + \omega_0^2 z = ae^{i\omega t}$$

Particular solution is a multiple of $ae^{i\omega t}$, so

$$z = Tae^{i\omega t}$$

Plug in and solve for T, and write it in polar form as

$$T = Ge^{-i\phi}$$

Then

$$z = \left(Ge^{-i\phi}\right)ae^{i\omega t}$$

$$= Gae^{i(\omega t - \phi)}$$

which gives

$$u = Ga\cos(\omega t - \phi)$$

To interpret, think of the DE as modeling a system with an input (the force applied) and an output (the resulting motion). Then we have

input : $a\cos\omega t$

output : $Ga\cos(\omega t - \phi)$

The system then introduces two changes

"gain" = factor change in amplitude $= G$

"phase shift" = shift of oscillation $= \phi$

Ex: Suppose $L(y) = y'' + 2cy' + \omega_0^2 y = a\cos\omega t$

and we solve the related complex equation

$$L(z) = ae^{i\omega t}$$

with

$$z = Tae^{i\omega t}.$$

and that $T = 2+5i$. What are the gain and the phase shift in the solution to the original equation?

Note we can write T in polar form as

$$T = Ge^{-i\phi}$$

$$G = \sqrt{29} \qquad -\phi = \arccos\left(\frac{2}{\sqrt{29}}\right)$$

Then $z = Tae^{i\omega t} = \sqrt{29}\, e^{i(-\phi)} ae^{i\omega t} = a\sqrt{29}\, e^{i(\omega t - \phi)}$

So $y = \mathrm{Re}(z) = a\sqrt{29}\cos(\omega t - \phi) = \sqrt{29}\, a\cos(\omega t - \phi)$

The gain is the factor of increase in the amplitude, which is $\boxed{\sqrt{29}}$.

The phase shift is ϕ, which is $\boxed{-\arccos\left(\frac{2}{\sqrt{29}}\right)}$.

Linear Transformations

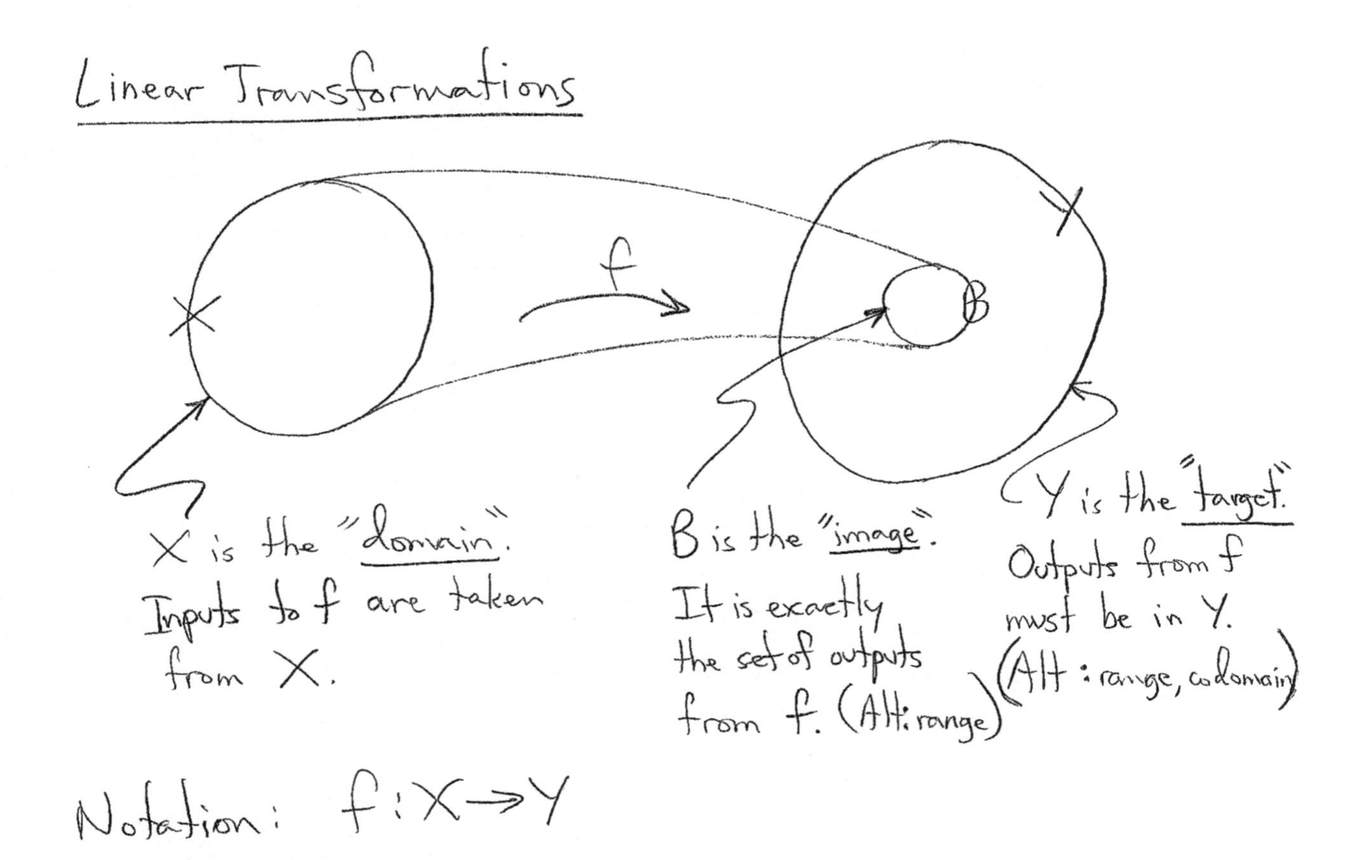

Notation: $f: X \to Y$

Ex: $f: \mathbb{R}^1 \to \mathbb{R}^1$, $f(x) = x^2$.

Note image is $[0, \infty)$, which is not the same as the target, $\mathbb{R}^1$.

Ex: $f: \mathbb{R}^2 \to \mathbb{R}^1$, $f(x,y) = x^4 + y^4 - 3x^2y - 4y^3 + x$

Hard to compute the image...
But easy to write simply $f: \mathbb{R}^2 \to \mathbb{R}^1$, and we might not need the image!

A linear transformation is a special kind of function.

Def: Given vector spaces V, W, a function

$$T : V \to W$$

is a <u>linear transformation</u> if,

for all $\vec{x}_1, \vec{x}_2 \in V$ and $c_1, c_2 \in \mathbb{R}$, we have

$$T(c_1\vec{x}_1 + c_2\vec{x}_2) = c_1T(\vec{x}_1) + c_2T(\vec{x}_2).$$

Alt: the linearity condition is equivalent to

$$T(c_1\vec{x}_1 + \ldots + c_n\vec{x}_n) = c_1T(\vec{x}_1) + \ldots + c_nT(\vec{x}_n)$$

and it is also equivalent to the pair

$$T(\vec{x}_1 + \vec{x}_2) = T(\vec{x}_1) + T(\vec{x}_2), \quad T(c\vec{x}) = cT(\vec{x})$$

You could say a linear transformation...

— "commutes with linear combinations"

or

— "commutes with addition and scalar multiplication".

Sometimes a process could have been represented as a function, but for whatever reason was not.

Ex: "A acts <u>linearly</u> on vectors by left multiplication."

$$\vec{x} \longrightarrow A\vec{x}$$

Ex: "It is <u>linear</u> to map points in $\mathbb{R}^3$ to their nearest points in the plane P through the origin."

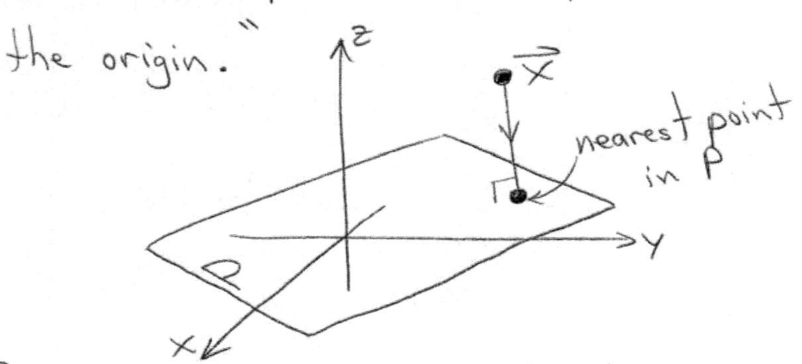

Ex: "$D^2 - e^x D$ is a <u>linear</u> operator."

$$f \longrightarrow (f'' - e^x f')$$

Ex: $a_{11}x_1 + \dots + a_{1n}x_n = b_1$
$\vdots$
$a_{m1}x_1 + \dots + a_{mn}x_n = b_m$

is a <u>linear</u> system of equations

$$\begin{pmatrix} x_1 \\ \vdots \\ x_n \end{pmatrix} \longrightarrow \begin{pmatrix} a_{11}x_1 + \dots + a_{1n}x_n \\ \vdots \\ a_{m1}x_1 + \dots + a_{mn}x_n \end{pmatrix}$$

Ex: Is $T: \mathbb{R}^2 \to \mathbb{R}^2$, defined by

$$T\begin{pmatrix} x \\ y \end{pmatrix} = \begin{pmatrix} x - y \\ 3x + 2y \end{pmatrix}$$

a l.t.?

Check:

$$T(a\vec{x}_1 + b\vec{x}_2) = T\left(a\begin{pmatrix} x_1 \\ y_1 \end{pmatrix} + b\begin{pmatrix} x_2 \\ y_2 \end{pmatrix}\right)$$

$$= T\begin{pmatrix} ax_1 + bx_2 \\ ay_1 + by_2 \end{pmatrix}$$

$$= \begin{pmatrix} (ax_1 + bx_2) - (ay_1 + by_2) \\ 3(ax_1 + bx_2) + 2(ay_1 + by_2) \end{pmatrix}$$

$$= \begin{pmatrix} ax_1 - ay_1 \quad + \quad bx_2 - by_2 \\ 3ax_1 + 2ay_1 \quad + \quad 3bx_2 + 2by_2 \end{pmatrix}$$

$$= a\begin{pmatrix} x_1 - y_1 \\ 3x_1 + 2y_1 \end{pmatrix} + b\begin{pmatrix} x_2 - y_2 \\ 3x_2 + 2y_2 \end{pmatrix}$$

$$= aT(\vec{x}_1) + bT(\vec{x}_2)$$

✓

Ex: Is $T: \mathbb{R}^2 \to \mathbb{R}^1$, defined by

$$T\begin{pmatrix} x \\ y \end{pmatrix} = x^2 + y$$

a l.t.?

No...

$$T\begin{pmatrix} 2 \\ 0 \end{pmatrix} = 4$$

$$2T\begin{pmatrix} 1 \\ 0 \end{pmatrix} = 2$$

(these are not equal)

Here are some interesting linear transformations:

① $D: D(a,b) \to F(a,b)$ defined by

$$D(f) = f'$$

This is a l.t. because

$$D(af_1 + bf_2) = (af_1 + bf_2)'$$

$$= af_1' + bf_2'$$

$$= aD(f_1) + bD(f_2) \quad \checkmark$$

② $T: \mathbb{R}^n \to \mathbb{R}^m$ defined by

$$T(\vec{x}) = A\vec{x}$$

(for a given $m \times n$ matrix A)

③ $S: C^{\infty}[a,b] \to \mathbb{R}^1$, defined by

$$S(f) = \int_a^b f(x)\,dx$$

④ $S_g: C^{\infty}[a,b] \to \mathbb{R}^1$, defined by

$$S_g(f) = \int_a^b f(x)\,g(x)\,dx$$

This is useful in statistics/probability, physics, ...

⑤ $\delta_0: C^{\infty} \to \mathbb{R}^1$, defined by

$$\delta_0(f) = f(0)$$

This is the "Dirac distribution" or "Dirac measure".

(Note, it is possible to relate this to ④ above with limits... Then it is tempting to call this a function, but this is problematic.)

⑥ $\delta_a^{[n]}: C^{\infty} \to \mathbb{R}^1$, defined by

$$\delta_a^{[n]}(f) = f^{[n]}(a)$$

⑦ In a linear DE $L(y) = g(x)$, the linear diff. op. L is a linear transformation

$$L: C^n \to C^0$$

Thm: The images of a basis for the domain completely determine all of the values of a l.t.

Pf: We consider a l.t. $T: V \to W$, and suppose $\{\vec{v}_1, \ldots, \vec{v}_k\}$ are a basis for V.

Then, for any $\vec{v} \in V$, we can write

$$\vec{v} = c_1\vec{v}_1 + \cdots + c_k\vec{v}_k$$

and then compute

$$T(\vec{v}) = T(c_1\vec{v}_1 + \cdots + c_k\vec{v}_k)$$
$$= c_1 T(\vec{v}_1) + \cdots + c_k T(\vec{v}_k)$$

Ex: Say we know that $T: \mathbb{R}^2 \to \mathbb{R}^3$ has

$$T\begin{pmatrix}1\\1\end{pmatrix} = \begin{pmatrix}1\\2\\3\end{pmatrix} \quad \text{and} \quad T\begin{pmatrix}1\\2\end{pmatrix} = \begin{pmatrix}2\\5\\7\end{pmatrix}$$

What is $T\begin{pmatrix}3\\4\end{pmatrix}$?

$$T\begin{pmatrix}3\\4\end{pmatrix} = T\left(2\begin{pmatrix}1\\1\end{pmatrix} + \begin{pmatrix}1\\2\end{pmatrix}\right)$$
$$= 2T\begin{pmatrix}1\\1\end{pmatrix} + T\begin{pmatrix}1\\2\end{pmatrix}$$
$$= 2\begin{pmatrix}1\\2\\3\end{pmatrix} + \begin{pmatrix}2\\5\\7\end{pmatrix} = \begin{pmatrix}4\\9\\13\end{pmatrix}$$

Def: The <u>kernel</u> of a linear transformation $T: V \to W$ is the subset of V defined by

$$\ker(T) = \{ v \in V \mid T(v) = 0 \}$$

Ex: Say $T: \mathbb{R}^n \to \mathbb{R}^m$ is defined by $T(\vec{x}) = A\vec{x}$

Then $\ker(T) = NS(A)$ = set of homogeneous sols.

Ex: Say $T(\vec{x}) = \begin{pmatrix} 1 & 0 & 0 \\ 0 & 1 & 0 \\ 0 & 0 & 0 \end{pmatrix} \vec{x}$, which projects vectors in $\mathbb{R}^3$ vertically to the xy-plane.

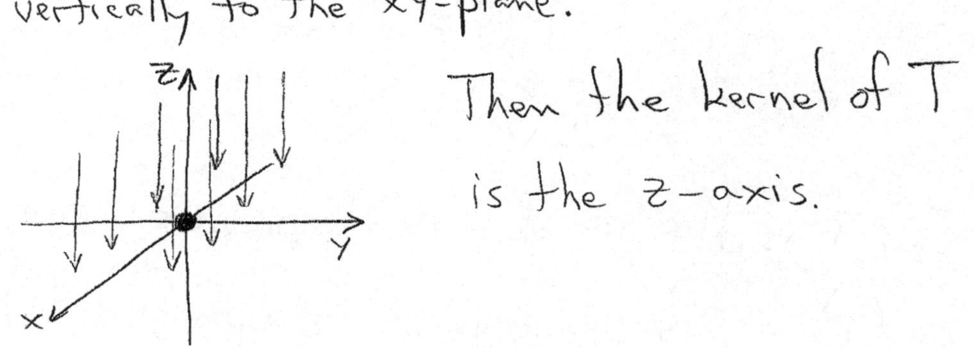

Then the kernel of T is the z-axis.

Ex: Say $L: C^n \to C^0$ is an nth order linear differential operator.

Then $\ker(L)$ is the set of solutions to the homogeneous equation

$$L(y) = 0$$

Def: The <u>image</u> of a linear transformation $T: V \to W$ is the subset of W defined by

$$\text{im}(T) = \{ \vec{w} \in W \mid \vec{w} = T(\vec{v}),\ \vec{v} \in V \}$$

Ex: Say $T(\vec{x}) = A\vec{x}$, with $A = \begin{pmatrix} | & & | \\ \vec{v}_1 & \cdots & \vec{v}_n \\ | & & | \end{pmatrix}$

Then $T(\vec{x}) = A\vec{x} = x_1\vec{v}_1 + \ldots + x_n\vec{v}_n$

So $\text{im}(T) = \{\text{all l.c.'s of cols of } A\}$
$= CS(A)$

Ex: Say T is the projection on the previous page.

$$\text{im}(T) = xy\text{-plane in } \mathbb{R}^3 = \text{span}\left\{ \begin{pmatrix} 1 \\ 0 \\ 0 \end{pmatrix}, \begin{pmatrix} 0 \\ 1 \\ 0 \end{pmatrix}, \begin{pmatrix} 0 \\ 0 \\ 0 \end{pmatrix} \right\}$$

Thm: Given a l.t. $T: V \to W$:

- $\ker(T)$ is a subspace of V

and

- $\text{im}(T)$ is a subspace of W

(Prove!)

Thm:) If V is a finite-dim. vectorspace, and $T: V \to W$ is a l.t., then

$$\dim(\ker(T)) + \dim(\mathrm{im}(T)) = \dim(V)$$

(Proved in book.)

Alt. proof:) We will see soon that every l.t. on a finite dim domain can be written as a matrix. So consider $T(\vec{x}) = A\vec{x}$, $T: \mathbb{R}^n \to \mathbb{R}^m$; we know

$$\dim(NS(A)) + \dim(CS(A)) = n$$

which confirms the equation above!

Cor.:) If $T: V \to W$ is a l.t., and is also one-to-one and onto (bijective), then

$$\dim(V) = \dim(W)$$

(Recall from single variable calculus that a function is "bijective" if it is a "perfect pairing" between the domain and the target. This is required for a function to be invertible.)

Back to LDE's...

How do we show that the set of solutions is n-dim.? ($L(y)=0$, nth order, g_i cont., $g_n \neq 0$.)

Observe:

① $S = \{\text{set of sols}\}$ is a vector space.

② $I = \{\text{set of initial values}\} = \mathbb{R}^n$ is also a v.s.

(an init. value is $\begin{array}{l} y(0)=k_1 \\ y'(0)=k_2 \\ \vdots \\ y^{[n-1]}(0)=k_n \end{array}$, rep. by $\begin{pmatrix} k_1 \\ \vdots \\ k_n \end{pmatrix} \in \mathbb{R}^n$)

③ $T: S \to I$, defined by $T(y) = \begin{pmatrix} y(0) \\ \vdots \\ y^{[n-1]}(0) \end{pmatrix}$

is a l.t. (check)

④ T is one-to-one because two solutions y_1, y_2 cannot have the same initial values.

(by the "uniqueness" in the exist-uniq. thm.)

⑤ T is onto because every init. value makes a sol.

(by the "existence" in the exist-uniq. thm.)

By previous theorem, $\dim(S) = \dim(I) = \dim(\mathbb{R}^n) = n$.

5.2 - Algebra of Linear Transformations

Linear transformations are functions. But, they can also be viewed as objects with operations — like numbers, or matrices.

<u>Def:</u> Say T, S are l.t.'s from $V \to W$. Then

$$(T+S)(\vec{v}) = T(\vec{v}) + S(\vec{v})$$

$$(kT)(\vec{v}) = kT(\vec{v})$$

Note that these are <u>also</u> <u>l.t.'s</u> from $V \to W$.

We can't conveniently multiply like this though (on outputs...)

But a product-like operation can be defined with <u>compositions</u>.

<u>Def:</u> Say $T: V \to W$ and $S: W \to U$ are l.t.'s.
Then $ST: V \to U$ is defined by

$$ST(\vec{v}) = (S \circ T)(\vec{v}) = S(T(\vec{v}))$$

Note that this is <u>also</u> a <u>l.t.</u>.

(Note: <u>target</u> of T must be the <u>domain</u> of S.)

With these operations (addition, scalar prod, composition),

we have an algebra on l.t.'s similar to matrix algebra:

Thm:) ① $S+T = T+S$
② $R+(S+T) = (R+S)+T$
③ $c(dT) = (cd)T$
④ $c(S+T) = cS + cT$
⑤ $(c+d)T = cT + dT$
⑥ $R(ST) = (RS)T$
⑦ $R(S+T) = RS + RT$
⑧ $(R+S)T = RT + ST$
⑨ $c(ST) = (cS)T = S(cT)$

Note of course that, just as with matrices, we do

not have composition commutativity:

$$ST \neq TS$$

To prove the above 9, note each side of each equation is a l.t.; prove they are the same by evaluating both on an arbitrary vector $\vec{v} \in V$.

Ex: To prove ⑧, we compute:

$$((R+S)T)(\vec{v}) = (R+S)(T(\vec{v})) = R(T(\vec{v})) + S(T(\vec{v}))$$

$$(RT+ST)(\vec{v}) = RT(\vec{v}) + ST(\vec{v}) = R(T(\vec{v})) + S(T(\vec{v}))$$

These are equal, as needed.

We can use this algebra to motivate some facts about solving DE's.

<u>Setup</u> ① Define $D: C^\infty \to C^\infty$ by

$$D(f) = f'$$

② Define $T_g: C^\infty \to C^\infty$ by

$$T_g(f) = gf$$

③ Observe that every LDE

$$g_n(x)\, y^{[n]}(x) + \cdots + g_0(x)\, y(x) = g(x)$$

has LHS that is a l.t. on y:

$$L(y) = (T_{g_n} D^n + \cdots + T_{g_0})(y) = g(x)$$

We can write this l.t. as

$$L = T_{q_n} D^n + \cdots + T_{q_0}$$

$$= q_n D^n + \cdots + q_0$$

④ Observe that solving a hom. LDE is equivalent to finding the kernel of the l.t. L.

⑤ If L has constant coefficients, then

$$L = p(D)$$

where p is the characteristic polynomial.

⑥ Note that LDO's do not commute:

$$DT_g(f) = D(gf) = f'g + fg'$$

$$T_g D(f) = T_g(f') = f'g$$

not equal!

But — CCLDO's do commute:

$$DT_{a_k}(f) = D(a_k f) = a_k f'$$

$$T_{a_k} D(f) = T_{a_k} f' = a_k f'$$

equal!

⑦ Note, CCLDO's factor just like their char. poly.:

$$p(\lambda) = (\lambda - r_1)^{m_1} \cdots (\lambda - r_k)^{m_k}$$

$$L = (D - r_1)^{m_1} \cdots (D - r_k)^{m_k}$$

Given these observations, we reconsider the question of solving a homogeneous CCLDE.

$$L(y) = 0$$

$$(D - r_1)^{m_1} \cdots (D - r_k)^{m_k} y = 0$$

Note that if r_i is one of the roots, the factors can be reordered as

$$(\ Q(D)\ )(D - r_i)\, y = 0$$

So, any y with $(D - r_1)y = 0$ would solve this.

That is:

$$y' = r_1 y$$

$$\Longrightarrow \boxed{y = e^{r_1 x}}$$

We can now prove our previous result about solutions to $L(y)=0$ when r has multiplicity m.

First, note that

$$(D-r)(x^k e^{rx})$$
$$= kx^{k-1}e^{rx} + x^k r e^{rx} - r x^k e^{rx}$$
$$= kx^{k-1}e^{rx}$$

when $k \geq 1$. (And $(D-r)(e^{rx}) = 0$.)

And we can rewrite $L = P(D) = Q(D)(D-r)^m$.

Consider one of the proposed solutions, $x^k e^{rx}$ $(0 \leq k \leq m-1)$.

Then

$$L(y) = Q(D)(D-r)^m (x^k e^{rx})$$
$$= Q(D)(D-r)^{m-k-1}(D-r)(D-r)^k (x^k e^{rx})$$
$$= Q(D)(D-r)^{m-k-1}(D-r)(k!\,e^{rx})$$
$$= Q(D)(D-r)^{m-k-1}(0)$$
$$= 0$$

We can also use this to motivate part of the big theorem about undetermined coefficients.

Here is an equivalent theorem.

Thm: Consider $L(y) = g(x) = Ax^k e^{rx}$ $\left(\begin{matrix} r = a+bi \\ k \in \mathbb{N} \end{matrix}\right)$

① If r is not a root of $p(\lambda)$, a particular solution is

$$y_p = (c_k x^k + \ldots + c_0) e^{rx}$$

② If r is a root of $p(\lambda)$ (of multiplicity m), a particular solution is

$$y_p = x^m (c_k x^k + \ldots + c_0) e^{rx}$$

(recall $e^{rx} = e^{ax+bix} = e^{ax}\cos bx + i e^{ax}\sin bx$, and can make arguments with linear combinations.)

Why the x^m? Well,

$$L(y_p) = Q(D)(D-r)^m \left(x^m (c_k x^k + \ldots + c_0) e^{rx}\right)$$
$$= Q(D) \left((d_k x^k + \ldots + d_0) e^{rx}\right) = g(x)$$

Part ① says we can solve for $d_k, \ldots, d_0$.

We can then back solve for $c_k, \ldots, c_0$

5.3 - Matrices and Change of Basis

We have already seen that:

Thm:) $\left(T(x) = A\vec{x}\right) \implies \left(T \text{ is a l.t. from } \mathbb{R}^n \to \mathbb{R}^m\right)$

In fact this works in both directions:

Thm:) If T is a l.t. from $\mathbb{R}^n \to \mathbb{R}^m$, then there exists an $m \times n$ matrix A with

$$T(\vec{x}) = A\vec{x}$$

Pf:) Let $\vec{a}_i = T(\vec{e}_i)$, and

$$A = \begin{pmatrix} | & & | \\ \vec{a}_1 & \cdots & \vec{a}_n \\ | & & | \end{pmatrix}$$

Then we simply check that

$$\begin{aligned} A\vec{x} &= x_1\vec{a}_1 + \cdots + x_n\vec{a}_n \\ &= x_1 T(\vec{e}_1) + \cdots + x_n T(\vec{e}_n) \\ &= T\left(x_1\vec{e}_1 + \cdots + x_n\vec{e}_n\right) \\ &= T(\vec{x}) \end{aligned}$$

Ex:) Note that rotations in $\mathbb{R}^2$ are l.t.'s; now we see that we can compute these rotations with matrices <u>and</u> that we can find this matrix by looking at the $\vec{e}_i$.

Let R_θ be rotation ccwise around $\vec{0}$ by angle θ.

Note $$R_\theta(\vec{e}_1) = \begin{pmatrix} \cos\theta \\ \sin\theta \end{pmatrix} \quad R_\theta(\vec{e}_2) = \begin{pmatrix} -\sin\theta \\ \cos\theta \end{pmatrix}$$

So $$R_\theta(\vec{x}) = A\vec{x}$$

with $$A = \begin{pmatrix} \cos\theta & -\sin\theta \\ \sin\theta & \cos\theta \end{pmatrix}$$

Note: The columns of A are the images by T of the standard basis vectors

Ex:) We know that for linear transformations $\mathbb{R}^n \to \mathbb{R}^m$

- composing l.t.'s gives l.t.'s
- we view this as a product on l.t.'s
- for each such l.t. there is a corresponding matrix

But, we do <u>not</u> yet know if the matrix for the product is the product of those matrices...

Q: What is the matrix for the composition of 2 l.t.'s?

$$\mathbb{R}^n \xrightarrow{T} \mathbb{R}^m \xrightarrow{S} \mathbb{R}^k$$

$$T(\vec{x}) = A\vec{x}\,, \quad S(\vec{y}) = B\vec{y} \qquad ST(\vec{x}) = C\vec{x}$$

We compute C one column at a time:

$$\vec{c_i} = ST(\vec{e_i}) = S(T(\vec{e_i})) = B(A(e_i))$$

$$= B\vec{a_i}$$

cannot use associativity — we have not proved that yet!

By matrix multiplication, this <u>is</u> the ith column of the product BA.

So the product we defined for l.t.'s corresponds nicely to matrix multiplication.

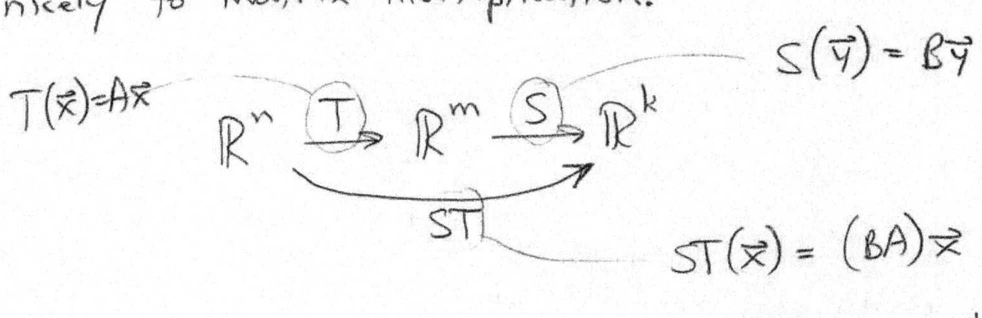

<u>Comments:</u> ① Proving associativity of matrix multiplication is now an immediate consequence of associativity of compositions.

$$R(ST) = (RS)T$$

$$\Updownarrow$$

$$C(BA) = (CB)A$$

② One can use this relationship between matrices and l.t.'s as the <u>definition</u> of matrices and operations ...

Motivates matrix addition, scalar mult., and matrix mult. ...

Bases and coordinates

Recall that if $\mathcal{V} = \{\vec{v}_1, \ldots, \vec{v}_n\}$ is a basis for V, then there is a <u>unique</u> way to write any $\vec{v} \in V$ as

$$\vec{v} = c_1\vec{v}_1 + \cdots + c_n\vec{v}_n$$

These unique coefficients are the "coordinates of $\vec{v}$ relative to this basis", and we write

$$[\vec{v}]_{\mathcal{V}} = \begin{pmatrix} c_1 \\ \vdots \\ c_n \end{pmatrix}$$

We can thus think of the basis $\mathcal{V}$ as giving a bijection between V (made up of vectors) and $\mathbb{R}^n$ (made up of coordinates relative to the basis).

Notation:

$$V \xleftrightarrow{\mathcal{V}} \mathbb{R}^n$$

$$\vec{v} \longrightarrow [\vec{v}]_{\mathcal{V}}$$

$$\begin{pmatrix} c_1 \\ \vdots \\ c_n \end{pmatrix}_{\mathcal{V}} \longleftarrow \begin{pmatrix} c_1 \\ \vdots \\ c_n \end{pmatrix}$$

You can check that this is itself a l. t..

Ex:) Consider P_2, the v.s. of polynomials of degree ≤ 2, and the basis $\mathcal{V} = \{\underbrace{x^2-1}_{\vec{v}_1}, \underbrace{2x-1}_{\vec{v}_2}, \underbrace{1}_{\vec{v}_3}\}$

We can write the polynomial $f = 3x^2 + 6x + 5$

as $f = 3\vec{v}_1 + 3\vec{v}_2 + 11\vec{v}_3$

Using the notation on the previous page then:

$$[f]_{\mathcal{V}} = \begin{pmatrix} 3 \\ 3 \\ 11 \end{pmatrix}$$

and

$$f = {}_{\mathcal{V}}\begin{pmatrix} 3 \\ 3 \\ 11 \end{pmatrix}$$

Ex:) $\mathbb{R}^3$ has a basis $\mathcal{V} = \left\{ \underbrace{\begin{pmatrix} 1 \\ 2 \\ 1 \end{pmatrix}}_{\vec{v}_1}, \underbrace{\begin{pmatrix} 3 \\ 2 \\ 4 \end{pmatrix}}_{\vec{v}_2}, \underbrace{\begin{pmatrix} 7 \\ 2 \\ 1 \end{pmatrix}}_{\vec{v}_3} \right\}$

We have $\begin{pmatrix} 5 \\ 6 \\ -3 \end{pmatrix} = 4\begin{pmatrix} 1 \\ 2 \\ 1 \end{pmatrix} - 2\begin{pmatrix} 3 \\ 2 \\ 4 \end{pmatrix} + 1\begin{pmatrix} 7 \\ 2 \\ 1 \end{pmatrix}$

So

$$\begin{pmatrix} 5 \\ 6 \\ -3 \end{pmatrix}_{\mathcal{V}} = \begin{pmatrix} 4 \\ -2 \\ 1 \end{pmatrix}$$

and

$$\begin{pmatrix} 5 \\ 6 \\ -3 \end{pmatrix} = {}_{\mathcal{V}}\begin{pmatrix} 4 \\ -2 \\ 1 \end{pmatrix}$$

Linear transformations

Using these ideas, we can use bases as a language for communicating about linear transformations.

Consider $T: V \to W$, with $\mathcal{V} = \{\vec{v}_1, \ldots, \vec{v}_n\}$ a basis for V and $\mathcal{W} = \{\vec{w}_1, \ldots, \vec{w}_m\}$ a basis for W:

$$\begin{array}{ccc} V & \xrightarrow{\quad T \quad} & W \\ \updownarrow \mathcal{V} & & \updownarrow \mathcal{W} \\ \mathbb{R}^n & & \mathbb{R}^m \end{array}$$

All of the arrows are l.t.'s, so the composition from $\mathbb{R}^n \to \mathbb{R}^m$ is a l.t. — and thus is rep'd by a matrix:

$$\begin{array}{ccc} V & \xrightarrow{\quad T \quad} & W \\ \updownarrow \mathcal{V} & & \updownarrow \mathcal{W} \\ \mathbb{R}^n & \xrightarrow{\quad A \quad} & \mathbb{R}^m \end{array}$$

We call this matrix A the <u>matrix of T w.r.t. the bases $\mathcal{V}$ and $\mathcal{W}$</u>, written $A = [T]_{\mathcal{V}}^{\mathcal{W}}$

$A = [T]_{\mathcal{V}}^{\mathcal{W}}$ "represents T" in the sense that it executes T in the language of the specified bases.

$$[T(\vec{v})]_{\mathcal{W}} = [T]_{\mathcal{V}}^{\mathcal{W}} [v]_{\mathcal{V}}$$

Ex:) Consider $D: P^3 \to P^2$ defined by $D(f) = f'$, and bases

$$\mathcal{V} = \{\underbrace{x^3}_{\vec{v}_1}, \underbrace{x^2}_{\vec{v}_2}, \underbrace{x}_{\vec{v}_3}, \underbrace{1}_{\vec{v}_4}\} \text{ for } P^3$$

$$\mathcal{W} = \{\underbrace{x^2}_{\vec{w}_1}, \underbrace{x}_{\vec{w}_2}, \underbrace{1}_{\vec{w}_3}\} \text{ for } P^2$$

Let's "sanity check" that

$$[D]_{\mathcal{V}}^{\mathcal{W}} = \begin{pmatrix} 3 & 0 & 0 & 0 \\ 0 & 2 & 0 & 0 \\ 0 & 0 & 1 & 0 \end{pmatrix}$$

by considering the vector $(x^2 + x)$ in P^3.

$$\begin{array}{ccc} P^3 & \xrightarrow{\quad D \quad} & P^2 \\ \updownarrow \mathcal{V} & & \updownarrow \mathcal{W} \\ \mathbb{R}^4 & \xrightarrow{[D]_{\mathcal{V}}^{\mathcal{W}}} & \mathbb{R}^3 \end{array}$$

First we go "over then down":

$$(x^2+x) \xrightarrow{D} (2x+1)$$

$$\updownarrow \mathcal{W}$$

$$\begin{pmatrix} 0 \\ 2 \\ 1 \end{pmatrix}$$

Alternatively, we go "down then over":

$$(x^2+x)$$

$$\updownarrow \mathcal{V}$$

$$\begin{pmatrix} 0 \\ 1 \\ 1 \\ 0 \end{pmatrix} \xrightarrow{[D]_{\mathcal{V}}^{\mathcal{W}}} \begin{pmatrix} 0 \\ 2 \\ 1 \end{pmatrix}$$

As required, we get the same result.

(Check other vectors also!)

How do we find the matrix $[T]_{\mathcal{V}}^{\mathcal{W}}$? We go "up and over" on our diagram:

$$\begin{array}{ccc} V & \xrightarrow{T} & W \\ \updownarrow \mathcal{V} & & \updownarrow \mathcal{W} \\ \mathbb{R}^n & \xrightarrow{[T]_{\mathcal{V}}^{\mathcal{W}}} & \mathbb{R}^m \end{array}$$

And recall that each <u>column</u> of $[T]_{\mathcal{V}}^{\mathcal{W}}$ is the image of the appropriate <u>standard basis vector</u>.

$$\begin{aligned} i\text{th col. of } [T]_{\mathcal{V}}^{\mathcal{W}} &= [T]_{\mathcal{V}}^{\mathcal{W}} \vec{e}_i \\ &= \left[T\left({}_{\mathcal{V}}[\vec{e}_i]\right)\right]_{\mathcal{W}} \quad \leftarrow \text{(up and over!)} \\ &= \left[T(\vec{v}_i)\right]_{\mathcal{W}} \end{aligned}$$

That is — the cols of $[T]_{\mathcal{V}}^{\mathcal{W}}$ are the $\mathcal{W}$ coordinates of the images of the corresponding $\mathcal{V}$ vectors.

Ex: For the previous example...

$$[D(\vec{v}_1)]_{\beta_W} = [D(x^3)]_{\beta_W} = [3x^2]_{\beta_W} = [3\vec{w}_1]_{\beta_W} = \begin{pmatrix} 3 \\ 0 \\ 0 \end{pmatrix}$$

$$[D(\vec{v}_2)]_{\beta_W} = [D(x^2)]_{\beta_W} = [2x]_{\beta_W} = [2\vec{w}_2]_{\beta_W} = \begin{pmatrix} 0 \\ 2 \\ 0 \end{pmatrix}$$

$$[D(\vec{v}_3)]_{\beta_W} = [D(x)]_{\beta_W} = [1]_{\beta_W} = [1\vec{w}_3]_{\beta_W} = \begin{pmatrix} 0 \\ 0 \\ 1 \end{pmatrix}$$

$$[D(\vec{v}_4)]_{\beta_W} = [D(1)]_{\beta_W} = [0]_{\beta_W} = \begin{pmatrix} 0 \\ 0 \\ 0 \end{pmatrix}$$

These columns are then assembled as

$$[D]_{\beta_V}^{\beta_W} = \begin{pmatrix} 3 & 0 & 0 & 0 \\ 0 & 2 & 0 & 0 \\ 0 & 0 & 1 & 0 \end{pmatrix}$$

confirming the previous assertion.

Ex:) Suppose we have

$$[T]_{\mathcal{V}}^{\mathcal{W}} = \begin{pmatrix} 1 & 3 & 0 \\ 0 & 1 & 1 \\ 1 & 2 & 4 \end{pmatrix}$$

How do we interpret this matrix as a statement about T?

Note, the columns are the $\mathcal{W}$ coordinates of the images of the $\mathcal{V}$ vectors... So:

$$T(\vec{v}_1) = 1\vec{w}_1 + 0\vec{w}_2 + 1\vec{w}_3$$

$$T(\vec{v}_2) = 3\vec{w}_1 + 1\vec{w}_2 + 2\vec{w}_3$$

$$T(\vec{v}_3) = 0\vec{w}_1 + 1\vec{w}_2 + 4\vec{w}_3$$

If the bases are the same, we get a similar conclusion.

Ex:) Suppose

$$[T]_{\mathcal{V}}^{\mathcal{V}} = \begin{pmatrix} 4 & 1 & 0 \\ 0 & 4 & 1 \\ 0 & 0 & 4 \end{pmatrix}$$

Then

$$T(\vec{v}_1) = 4\vec{v}_1$$

$$T(\vec{v}_2) = 1\vec{v}_1 + 4\vec{v}_2$$

$$T(\vec{v}_3) = 1\vec{v}_2 + 4\vec{v}_3$$

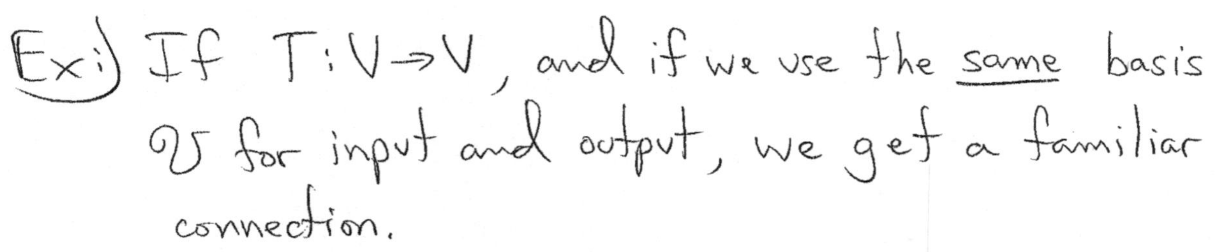

Ex:) If $T: V \to V$, and if we use the <u>same</u> basis $\mathcal{V}$ for input and output, we get a familiar connection.

$$\text{ith col. of } [T]_{\mathcal{V}}^{\mathcal{V}} = [T(\vec{v}_i)]_{\mathcal{V}}$$

That is — <u>the columns of the matrix</u> are the <u>images of the basis vectors</u> ($\mathcal{V}$, and in terms of $\mathcal{V}$; instead of $\mathcal{S}$.)

Ex:) Suppose we wish to describe the l.t. that flips over the line $X = 2Y$. Consider the basis

$$\mathcal{V} = \{\vec{v}_1, \vec{v}_2\}, \quad [\vec{v}_1]_{\mathcal{S}} = \begin{pmatrix} 2 \\ 1 \end{pmatrix}, \quad [\vec{v}_2]_{\mathcal{S}} = \begin{pmatrix} -1 \\ 2 \end{pmatrix}$$

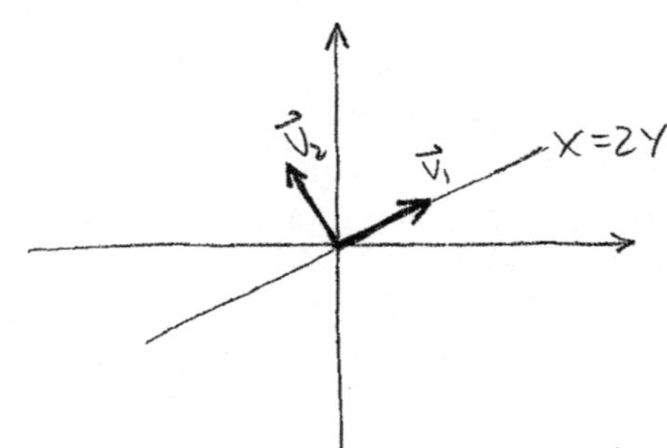

We easily observe

$$T(\vec{v}_1) = \vec{v}_1$$
$$T(\vec{v}_2) = -\vec{v}_2$$

Then $[T(\vec{v}_1)]_{\mathcal{V}} = \begin{pmatrix} 1 \\ 0 \end{pmatrix}$ and $[T(\vec{v}_2)]_{\mathcal{V}} = \begin{pmatrix} 0 \\ -1 \end{pmatrix}$

So $[T]_{\mathcal{V}}^{\mathcal{V}} = \begin{pmatrix} 1 & 0 \\ 0 & -1 \end{pmatrix}$.

Ex: Say $T : V \to W$

bases $\mathcal{S}_1 = \{\vec{e}_1, \ldots, \vec{e}_n\}$, $\mathcal{V} = \{\vec{v}_1, \ldots, \vec{v}_n\}$ (bases of V)

bases $\mathcal{S}_2 = \{\vec{e}_1, \ldots, \vec{e}_m\}$, $\mathcal{W} = \{\vec{w}_1, \ldots, \vec{w}_m\}$ (bases of W)

Let $[\vec{v}_1]_{\mathcal{S}_1} = \begin{pmatrix} c_1 \\ \vdots \\ c_n \end{pmatrix}$ $\quad [\vec{w}_1]_{\mathcal{S}_2} = \begin{pmatrix} d_1 \\ \vdots \\ d_m \end{pmatrix}$

$$[T]_{\mathcal{S}_1}^{\mathcal{S}_2} = A$$

$$[T]_{\mathcal{V}}^{\mathcal{W}} = M$$

Suppose it happens that $T(\vec{v}_1) = \vec{w}_1$. Then the following all make that same statement:

$$A \begin{pmatrix} c_1 \\ \vdots \\ c_n \end{pmatrix} = \begin{pmatrix} d_1 \\ \vdots \\ d_m \end{pmatrix}$$

$$[T]_{\mathcal{S}_1}^{\mathcal{S}_2} [\vec{v}_1]_{\mathcal{S}_1} = [\vec{w}_1]_{\mathcal{S}_2}$$

$$T(\vec{v}_1) = \vec{w}_1$$

$$[T]_{\mathcal{V}}^{\mathcal{W}} [\vec{v}_1]_{\mathcal{V}} = [\vec{w}_1]_{\mathcal{W}}$$

$$M \begin{pmatrix} 1 \\ 0 \\ \vdots \\ 0 \end{pmatrix} = \begin{pmatrix} 1 \\ 0 \\ \vdots \\ 0 \end{pmatrix}$$

What you know about bases used for some items in an equation gives you information on how to interpret other items in that equation.

Change of Basis (for vectors)

Suppose we know $[\vec{v}]_{\mathcal{V}}$ and we want to find $[\vec{v}]_{\mathcal{W}}$.

We call this a <u>change of basis</u>.

Note that this is accomplished by the identity transformation, with these bases used for input, output:

$$\begin{array}{ccc} V & \xrightarrow{\quad I \quad} & V \\ \updownarrow \mathcal{V} & & \updownarrow \mathcal{W} \\ \mathbb{R}^n & \xrightarrow{\ [I]_{\mathcal{V}}^{\mathcal{W}}\ } & \mathbb{R}^n \end{array}$$

So we have

(Change of basis matrix)

$$[\vec{v}]_{\mathcal{W}} = [I]_{\mathcal{V}}^{\mathcal{W}} [\vec{v}]_{\mathcal{V}}$$

As before, the <u>columns of $[I]$</u> are the <u>images of</u> the basis vectors:

$$\text{ith col. of } [I]_{\mathcal{V}}^{\mathcal{W}} = [I(\vec{v}_i)]_{\mathcal{W}}$$

$$= [\vec{v}_i]_{\mathcal{W}}$$

In this case then, we have that the columns are the $\mathcal{V}$ vectors written in the $\mathcal{W}$ basis.

Ex: Let $\mathcal{V} = \{\vec{v}_1, \vec{v}_2\}$, $\mathcal{S} = \{\vec{e}_1, \vec{e}_2\}$, with

$$[\vec{v}_1]_{\mathcal{S}} = \begin{pmatrix} 1 \\ 1 \end{pmatrix} \quad , \quad [\vec{v}_2]_{\mathcal{S}} = \begin{pmatrix} -1 \\ 1 \end{pmatrix}$$

What is the change of basis matrix from $\mathcal{V}$ to $\mathcal{S}$?

From previous discussion,

$$[I]_{\mathcal{V}}^{\mathcal{S}} = \left([\vec{v}_1]_{\mathcal{S}} \;\middle|\; [\vec{v}_2]_{\mathcal{S}} \right) = \begin{pmatrix} 1 & -1 \\ 1 & 1 \end{pmatrix}$$

Sanity check:

$$\underbrace{\begin{pmatrix} 1 & -1 \\ 1 & 1 \end{pmatrix}}_{[I]_{\mathcal{V}}^{\mathcal{S}}} \underbrace{\begin{pmatrix} 1 \\ 0 \end{pmatrix}}_{[\vec{v}_1]_{\mathcal{V}}} = \underbrace{\begin{pmatrix} 1 \\ 1 \end{pmatrix}}_{[\vec{v}_1]_{\mathcal{S}}}$$

and

$$\underbrace{\begin{pmatrix} 1 & -1 \\ 1 & 1 \end{pmatrix}}_{[I]_{\mathcal{V}}^{\mathcal{S}}} \underbrace{\begin{pmatrix} 0 \\ 1 \end{pmatrix}}_{[\vec{v}_2]_{\mathcal{V}}} = \underbrace{\begin{pmatrix} -1 \\ 1 \end{pmatrix}}_{[\vec{v}_2]_{\mathcal{S}}}$$

Note: $\left([I]_{\mathcal{V}}^{\mathcal{W}}\right)^{-1} = [I]_{\mathcal{W}}^{\mathcal{V}}$

because

$$[I]_{\mathcal{V}}^{\mathcal{W}} [I]_{\mathcal{W}}^{\mathcal{V}} = I$$

In this course, we will be using the change of basis matrix like this:

$$\underbrace{\begin{bmatrix} | & & | \\ [\vec{v}_1]_{\mathcal{W}} & \cdots & [\vec{v}_n]_{\mathcal{W}} \\ | & & | \end{bmatrix}}_{[I]_{\mathcal{V}}^{\mathcal{W}}} [\vec{x}]_{\mathcal{V}} = [\vec{x}]_{\mathcal{W}}$$

So we will refer to it as "C.O.B. matrix <u>from $\mathcal{V}$ to $\mathcal{W}$</u>".

But note also...

$$\begin{bmatrix} | & & | \\ \vec{w}_1 & \cdots & \vec{w}_n \\ | & & | \end{bmatrix} \overbrace{\begin{bmatrix} | & & | \\ [\vec{v}_1]_{\mathcal{W}} & \cdots & [\vec{v}_n]_{\mathcal{W}} \\ | & & | \end{bmatrix}}^{[I]_{\mathcal{V}}^{\mathcal{W}}} = \begin{bmatrix} | & & | \\ \vec{v}_1 & \cdots & \vec{v}_n \\ | & & | \end{bmatrix}$$

Thus in some contexts (including the textbook), it is referred to as "C.O.B. matrix <u>from $\mathcal{W}$ to $\mathcal{V}$</u>".

Be sure to interpret the terminology appropriately to the context!

Change of Basis (for linear transformations)

Suppose we have a l.t. $T: V \to W$, bases $\mathcal{V}, \mathcal{V}'$ for V, and bases $\mathcal{W}, \mathcal{W}'$ for W.

Suppose we know $[T]_{\mathcal{V}}^{\mathcal{W}}$. How do we find $[T]_{\mathcal{V}'}^{\mathcal{W}'}$?

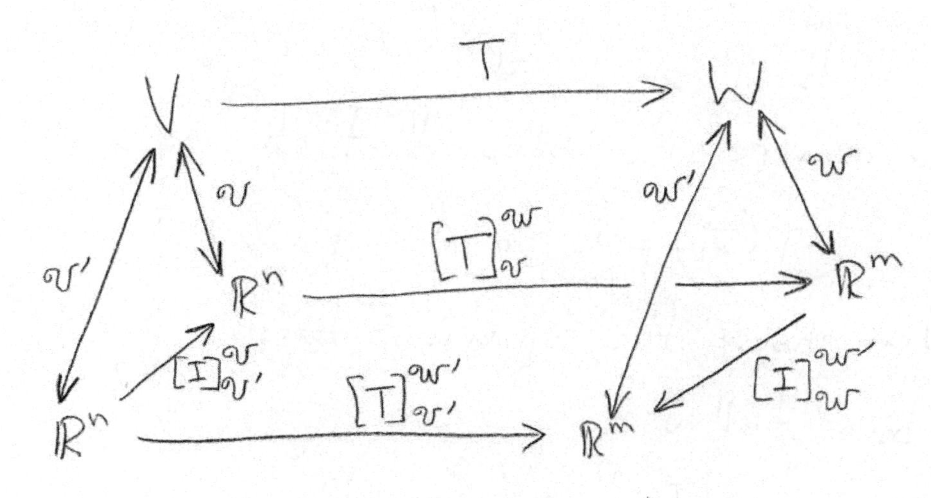

We see from the above diagram that

$$[T]_{\mathcal{V}'}^{\mathcal{W}'} = [I]_{\mathcal{W}}^{\mathcal{W}'} [T]_{\mathcal{V}}^{\mathcal{W}} [I]_{\mathcal{V}'}^{\mathcal{V}}$$

Ex: Suppose we want to find the matrix A with

$$A\begin{pmatrix}1\\3\end{pmatrix}=\begin{pmatrix}2\\5\end{pmatrix} \quad \text{and} \quad A\begin{pmatrix}2\\5\end{pmatrix}=\begin{pmatrix}1\\3\end{pmatrix}$$

Let's choose to view these as statements w.r.t. $\mathcal{S}$:

$$[T]_{\mathcal{S}}^{\mathcal{S}}[\vec{v}_1]_{\mathcal{S}}=[\vec{v}_2]_{\mathcal{S}} \text{ and } [T]_{\mathcal{S}}^{\mathcal{S}}[\vec{v}_2]_{\mathcal{S}}=[\vec{v}_1]_{\mathcal{S}}$$

with $[T]_{\mathcal{S}}^{\mathcal{S}}=A$, $[\vec{v}_1]_{\mathcal{S}}=\begin{pmatrix}1\\3\end{pmatrix}$, $[\vec{v}_2]_{\mathcal{S}}=\begin{pmatrix}2\\5\end{pmatrix}$

So we can rewrite independent of basis as

$$T(\vec{v}_1)=\vec{v}_2 \quad \text{and} \quad T(\vec{v}_2)=\vec{v}_1$$

Let's then choose the convenient basis $\mathcal{V}=\{\vec{v}_1,\vec{v}_2\}$; the above tell us

$$[T]_{\mathcal{V}}^{\mathcal{V}}=\begin{pmatrix}0&1\\1&0\end{pmatrix}$$

We also know

$$[I]_{\mathcal{V}}^{\mathcal{S}}=\begin{pmatrix}1&2\\3&5\end{pmatrix} \quad , \quad [I]_{\mathcal{S}}^{\mathcal{V}}=\begin{pmatrix}-5&2\\3&-1\end{pmatrix}$$

Then

$$\begin{aligned} A=[T]_{\mathcal{S}}^{\mathcal{S}} &= [I]_{\mathcal{V}}^{\mathcal{S}}[T]_{\mathcal{V}}^{\mathcal{V}}[I]_{\mathcal{S}}^{\mathcal{V}} \\ &= \begin{pmatrix}1&2\\3&5\end{pmatrix}\begin{pmatrix}0&1\\1&0\end{pmatrix}\begin{pmatrix}-5&2\\3&-1\end{pmatrix} \\ &= \begin{pmatrix}-7&3\\-16&7\end{pmatrix} \end{aligned}$$

Ex: Suppose we want to find the matrix B with

$$B\begin{pmatrix}1\\3\end{pmatrix}=\begin{pmatrix}3\\2\end{pmatrix} \quad \text{and} \quad B\begin{pmatrix}2\\5\end{pmatrix}=\begin{pmatrix}4\\3\end{pmatrix}$$

Again we choose to view these as statements w.r.t. $\mathcal{S}$:

$$[S]_{\mathcal{S}}^{\mathcal{S}}[\vec{v}_1]_{\mathcal{S}}=[\vec{w}_1]_{\mathcal{S}} \quad \text{and} \quad [S]_{\mathcal{S}}^{\mathcal{S}}[\vec{v}_2]_{\mathcal{S}}=[\vec{w}_2]_{\mathcal{S}}$$

with $[S]_{\mathcal{S}}^{\mathcal{S}}=B$, $[\vec{v}_1]_{\mathcal{S}}=\begin{pmatrix}1\\3\end{pmatrix}$, $[\vec{v}_2]_{\mathcal{S}}=\begin{pmatrix}2\\5\end{pmatrix}$, $[\vec{w}_1]_{\mathcal{S}}=\begin{pmatrix}3\\2\end{pmatrix}$, $[\vec{w}_2]_{\mathcal{S}}=\begin{pmatrix}4\\3\end{pmatrix}$

So we can rewrite our givens as

$$S(\vec{v}_1)=\vec{w}_1 \quad \text{and} \quad S(\vec{v}_2)=\vec{w}_2$$

We choose the convenient bases $\mathcal{V}=\{\vec{v}_1,\vec{v}_2\}$ (for the inputs)
$\mathcal{W}=\{\vec{w}_1,\vec{w}_2\}$ (for the outputs)

and can conclude

$$[S]_{\mathcal{V}}^{\mathcal{W}}=\begin{pmatrix}1&0\\0&1\end{pmatrix}$$

We also know

$$[I]_{\mathcal{W}}^{\mathcal{S}}=\begin{pmatrix}3&4\\2&3\end{pmatrix}, \quad [I]_{\mathcal{S}}^{\mathcal{V}}=\begin{pmatrix}-5&2\\3&-1\end{pmatrix}$$

Then

$$B=[S]_{\mathcal{S}}^{\mathcal{S}}=[I]_{\mathcal{W}}^{\mathcal{S}}\,[S]_{\mathcal{V}}^{\mathcal{W}}\,[I]_{\mathcal{S}}^{\mathcal{V}}$$

$$=\begin{pmatrix}3&4\\2&3\end{pmatrix}\begin{pmatrix}1&0\\0&1\end{pmatrix}\begin{pmatrix}-5&2\\3&-1\end{pmatrix}$$

$$=\begin{pmatrix}-3&2\\-1&1\end{pmatrix}$$

Eigenvectors and Eigenvalues

Often, a matrix A will have special (nonzero) vectors whose images are scalar multiples of themselves:

$$\text{(square)} \rightarrow A\vec{v} = \lambda\vec{v} \quad \lambda \leftarrow \text{(scalar!)}, \quad \vec{v} \leftarrow (\vec{v} \neq \vec{0})$$

This $\vec{v}$ is an <u>eigenvector</u>, and the λ is the <u>eigenvalue</u>.

(Similarly for linear transformations, and $\vec{v}$ could represent other things — function (eigenfunction), quantum state (eigenstate), ...)

How do we find them? Note

$$A\vec{v} = \lambda\vec{v} \quad \Longleftrightarrow \quad (A - \lambda I)\vec{v} = \vec{0}$$

So an eigenvalue λ makes $(A - \lambda I)$ <u>singular</u>.

Equivalently, $\quad \det(A - \lambda I) = 0$

This is a polynomial in λ, called the <u>characteristic polynomial</u>.

Note, the eigenvalues are simply the roots of the characteristic polynomial

Ex!) What are the eigenvalues & eigenvectors of

$$A = \begin{pmatrix} 1 & -3 \\ -2 & 2 \end{pmatrix}$$

The char. poly. is

$$\det(A-\lambda I) = \det\begin{pmatrix} 1-\lambda & -3 \\ -2 & 2-\lambda \end{pmatrix}$$

$$= (1-\lambda)(2-\lambda) - (-2)(-3)$$

$$= \lambda^2 - 3\lambda - 4$$

$$= (\lambda+1)(\lambda-4)$$

The roots, and thus the eigenvalues, are $-1, 4$.

To find the eigenvectors, we look for solutions to $(A-\lambda I)\vec{v} = \vec{0}$:

For $\lambda = -1$:

$$(A-\lambda I)\vec{v} = \begin{pmatrix} 2 & -3 \\ -2 & 3 \end{pmatrix}\vec{v} = \vec{0}$$

$$\ldots \Rightarrow \vec{v} = k\begin{pmatrix} 3 \\ 2 \end{pmatrix}$$

Convenient trick for solving some systems:

Consider from the previous calculation

$$(A-\lambda I)\vec{v} = \begin{pmatrix} 2 & -3 \\ -2 & 3 \end{pmatrix}\vec{v} = \vec{0}$$

What is the rank of this matrix?

- It can't be 2, because we know it is singular.
- It can't be 0, because the matrix is nonzero.

$\Rightarrow$ rank $(A-\lambda I) = 1$.

By the rank-nullity theorem then, $\dim(NS) = 2-1 = 1$

So the set of solutions we are looking for is 1-dim, and that means we need only find <u>any solution</u>, and it will be a basis!

Easy to see $\begin{pmatrix} 3 \\ 2 \end{pmatrix}$ is a solution, so the entire nullspace is

$$\vec{v} = k\begin{pmatrix} 3 \\ 2 \end{pmatrix}$$

For $\lambda = 4$:

$$(A - \lambda I)\vec{v} = \begin{pmatrix} -3 & -3 \\ -2 & -2 \end{pmatrix} \vec{v} = \vec{0}$$

$$\ldots \Rightarrow \vec{v} = k \begin{pmatrix} 1 \\ -1 \end{pmatrix}$$

In this example, we had a $\underline{2} \times \underline{2}$ matrix, a degree $\underline{2}$ char. poly., $\underline{2}$ distinct roots/eigenvalues, and $\underline{2}$ eigenvectors (l.i.).

It's great when this happens... unfortunately this is not always the case.

Ex! Consider

$$A = \begin{pmatrix} 1 & 1 \\ 0 & 1 \end{pmatrix}$$

Char. poly is

$$(1-\lambda)(1-\lambda) - (1)(0) = (1-\lambda)^2$$

$$\Rightarrow \lambda = 1$$

Eigenvectors are solutions to

$$(A - \lambda I)\vec{v} = \vec{0}$$

$$\begin{pmatrix} 0 & 1 \\ 0 & 0 \end{pmatrix} \vec{v} = \vec{0}$$

$$\ldots \Rightarrow \vec{v} = k \begin{pmatrix} 1 \\ 0 \end{pmatrix}$$

There were 2 roots if you count multiplicity... but, in this case there is only 1 eigenvector.

Ex: Let

$$A = \begin{pmatrix} 2 & -1 & 3 \\ 0 & -1 & 0 \\ 0 & 0 & -1 \end{pmatrix}$$

Char. poly is $\det(\lambda I - A) = \cdots = (\lambda-2)(\lambda+1)^2$

Eigen values are $2, -1$.

For $\lambda = 2$:

$$(A-\lambda I)\vec{v} = \begin{pmatrix} 0 & -1 & 3 \\ 0 & -3 & 0 \\ 0 & 0 & -3 \end{pmatrix} \vec{v} = \vec{0}$$

$$\cdots \Rightarrow \begin{pmatrix} 0 & 1 & 0 \\ 0 & 0 & 1 \\ 0 & 0 & 0 \end{pmatrix} \vec{v} = \vec{0}$$

$$\Rightarrow \vec{v} = k \begin{pmatrix} 1 \\ 0 \\ 0 \end{pmatrix}$$

For $\lambda = -1$:

$$(A-\lambda I)\vec{v} = \begin{pmatrix} 3 & -1 & 3 \\ 0 & 0 & 0 \\ 0 & 0 & 0 \end{pmatrix} \vec{v} = \vec{0}$$

$$\dots \Rightarrow \vec{v} = k_1 \begin{pmatrix} 1 \\ 3 \\ 0 \end{pmatrix} + k_2 \begin{pmatrix} -1 \\ 0 \\ 1 \end{pmatrix}$$

Here we get 2 (l.i.) eigenvectors for the eigenvalue of multiplicity 2. In total we have 3 eigenvalues (with multiplicity) and 3 eigenvectors (l.i.).

<u>Note</u>: ① eigenvalues are roots of $\det(A-\lambda I)$

② eigenvectors are solutions to $(A-\lambda I)\vec{v} = \vec{0}$

.... in $NS(A-\lambda I)$

We define the <u>eigenspace</u> for the eigenvalue λ by

$$E_\lambda = NS(A-\lambda I)$$

Ex: In previous example:

E_2 has basis $\left\{ \begin{pmatrix} 1 \\ 0 \\ 0 \end{pmatrix} \right\}$

E_{-1} has basis $\left\{ \begin{pmatrix} 1 \\ 3 \\ 0 \end{pmatrix}, \begin{pmatrix} -1 \\ 0 \\ 1 \end{pmatrix} \right\}$

Thm: If λ is a root of multiplicity m, then

$$1 \leq \dim(E_\lambda) \leq m$$

(We will not prove this)

So the most (l.i.) eigenvectors possible is when one has equality in the above for all λ, in which case

$$\begin{array}{ccccc} \dim(E_{\lambda_1}) & + & \cdots & + & \dim(E_{\lambda_k}) \\ \| & & & & \| \\ m_1 & + & \cdots & + & m_k \end{array} = n$$

Ex: In previous example,

$\lambda = 2$ had multiplicity $\underline{1}$, and $\dim(E_2) = \underline{1}$

$\lambda = -1$ had multiplicity $\underline{2}$, and $\dim(E_{-1}) = \underline{2}$

In these cases, the l.t. has a simple interpretation w.r.t. the eigenvectors.

Geometrically:

Ex! $A = \begin{pmatrix} 1 & -3 \\ -2 & 2 \end{pmatrix}$, $\lambda_1 = -1$, $\vec{v}_1 = \begin{pmatrix} 3 \\ 2 \end{pmatrix}$

$\lambda_2 = 4$, $\vec{v}_2 = \begin{pmatrix} 1 \\ -1 \end{pmatrix}$

Ex! $A = \begin{pmatrix} 2 & -1 & 3 \\ 0 & -1 & 0 \\ 0 & 0 & 1 \end{pmatrix}$ $\quad \lambda_1 = 2 \quad \vec{v}_1 = \begin{pmatrix} 1 \\ 0 \\ 0 \end{pmatrix}$

$\lambda_2 = -1 \quad \vec{v}_{21} = \begin{pmatrix} 1 \\ 3 \\ 0 \end{pmatrix} \quad \vec{v}_{22} = \begin{pmatrix} -1 \\ 0 \\ 1 \end{pmatrix}$

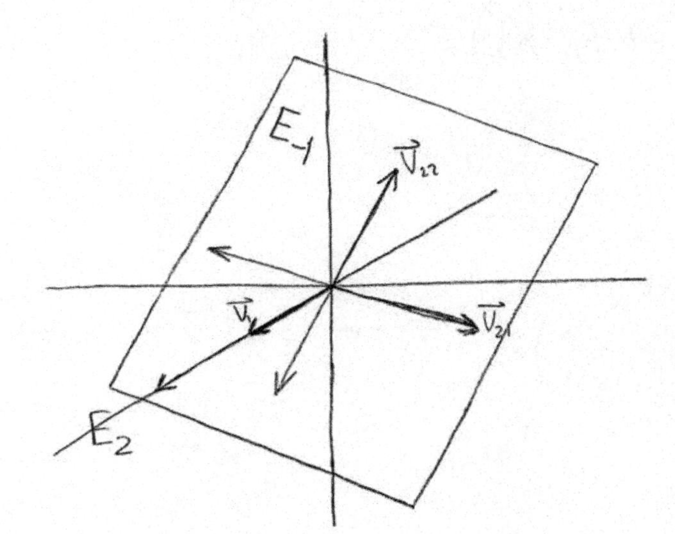

Note, these matrices will be simple w.r.t. the basis of evects.

Note that our char. poly is a poly. with real coefficients. But, it might have complex roots...

Ex:) Consider the rotation ($\pi/2$ ccwise) given by

$$A = \begin{pmatrix} 0 & -1 \\ 1 & 0 \end{pmatrix}$$

Geometrically it would seem impossible for this to have eigenvectors... but:

$$\det(\lambda I - A) = \lambda^2 + 1 = 0$$

$$\Rightarrow \lambda = i, -i$$

For $\lambda = i$: $(A - \lambda I)\vec{v} = \vec{0}$

$$\begin{pmatrix} -i & -1 \\ 1 & -i \end{pmatrix} \vec{v} = \vec{0}$$

$$\vdots$$

$$\begin{pmatrix} 1 & -i \\ 0 & 0 \end{pmatrix} \vec{v} = \vec{0} \quad \Rightarrow \vec{v}_1 = \begin{pmatrix} 1 \\ -i \end{pmatrix}$$

For $\lambda = -i$:

$$\begin{pmatrix} i & -1 \\ 1 & i \end{pmatrix} \vec{v} = \vec{0}$$

$$\vdots$$

$$\begin{pmatrix} 1 & i \\ 0 & 0 \end{pmatrix} \vec{v} = 0 \quad \Rightarrow \quad \vec{v}_2 = \begin{pmatrix} 1 \\ i \end{pmatrix}$$

So the rotation <u>matrix</u> has <u>complex</u> eigenvectors.

Thm: If A is real with complex eigenvalue λ, corresponding eigenspace E_λ with basis $\{\vec{v}_1, \dots, \vec{v}_k\}$, then $\bar{\lambda}$ is an eigenvalue, and $\{\overline{\vec{v}_1}, \dots, \overline{\vec{v}_k}\}$ is a basis for $E_{\bar{\lambda}}$.

(Proved in book.)

5.5 – Diagonalization and Jordan Canonical Form

In this section we will see strong connections between change of basis and eigenvectors.

Def:) A, B are <u>similar</u> if there is an invertible P with

$$B = P^{-1}AP$$

(here we say we are "conjugating" A by P; and A, B are "conjugates" of each other.)

Note, we have seen this sort of equation before...

$$[T]_{\mathcal{V}'}^{\mathcal{V}'} = \left([I]_{\mathcal{V}'}^{\mathcal{V}}\right)^{-1} [T]_{\mathcal{V}}^{\mathcal{V}} [I]_{\mathcal{V}'}^{\mathcal{V}}$$

In fact, all similar matrices can be viewed this way, because every invertible matrix P is a change of basis matrix.

Thm:) If P is invertible, then there exists a basis $\mathcal{V}$ for which $P = [I]_{\mathcal{V}}^{\mathcal{S}}$.

Pf:) The columns of P are this basis. (Note, because P is invertible, its columns are independent and thus form a basis.)

This theorem allows us to make a nice interpretation of this next definition.

Def:) A matrix is <u>diagonalizable</u> if it is similar to a diagonal matrix.

In light of the previous theorem, we can rephrase this as

Thm:) Say $A = [T]_{\mathcal{S}}^{\mathcal{S}}$. Then A is diagonalizable iff there exists a basis $\mathcal{V}$ for which

$$[T]_{\mathcal{V}}^{\mathcal{V}}$$

is a diaginal matrix.

Diagonal matrices are useful for many reasons. One of these is that the eigenvalues and eigenvectors are obvious.

Thm:) If D is diagonal with ith diagonal entry equal to d_i, then the eigenvalues are $\{d_i\}$ and the eigenvectors are $\vec{e}_i$.

Similarly then, if A is diagonalizable, then the basis vectors in $\mathcal{V}$ are eigenvectors.

Sometimes we say a matrix can be "put into a form" (say diagonal, or rref). What does this mean? If you are changing (clearly!) the matrix, how is "it" being "put into a form"? Is "it" still "it"?

<u>Something is being preserved</u>!

	row reduction	diagonalization	
Structure	- incr. lead zeroes - pivs = 1 - piv. cols otherwise 0's	- off diag = 0	
Process	row operations aka reversible operations aka elementary matrices	conjugation aka change of basis	
Preserved	solution set	linear transformation	
Possible?	Always	?	

Ex:) Recall that

$$A = \begin{pmatrix} 1 & -3 \\ -2 & 2 \end{pmatrix}$$

has eigenvalues and eigenvectors

$$\lambda_1 = -1 \qquad \vec{v}_1 = \begin{pmatrix} 3 \\ 2 \end{pmatrix}$$

$$\lambda_2 = 4 \qquad \vec{v}_2 = \begin{pmatrix} 1 \\ -1 \end{pmatrix}$$

What happens if we form a basis $\mathcal{V} = \{\vec{v}_1, \vec{v}_2\}$ and write $A = [T]_{\mathcal{S}}^{\mathcal{S}}$ in this basis? Well,

$[T]_{\mathcal{V}}^{\mathcal{V}}$ has columns: ① $[T(\vec{v}_1)]_{\mathcal{V}} = [-\vec{v}_1]_{\mathcal{V}} = \begin{pmatrix} -1 \\ 0 \end{pmatrix}$

② $[T(\vec{v}_2)]_{\mathcal{V}} = [4\vec{v}_2]_{\mathcal{V}} = \begin{pmatrix} 0 \\ 4 \end{pmatrix}$

So $[T]_{\mathcal{V}}^{\mathcal{V}} = \begin{pmatrix} -1 & 0 \\ 0 & 4 \end{pmatrix}$

The change of basis connection is that

$$[T]_{\mathcal{V}}^{\mathcal{V}} = \left([I]_{\mathcal{V}}^{\mathcal{S}}\right)^{-1} [T]_{\mathcal{S}}^{\mathcal{S}} [I]_{\mathcal{V}}^{\mathcal{S}}$$

$$\begin{pmatrix} -1 & 0 \\ 0 & 4 \end{pmatrix} = \begin{pmatrix} 3 & 1 \\ 2 & -1 \end{pmatrix}^{-1} \begin{pmatrix} 1 & -3 \\ -2 & 2 \end{pmatrix} \begin{pmatrix} 3 & 1 \\ 2 & -1 \end{pmatrix}$$

eigenvalues (pointing to the diagonal entries -1 and 4) eigenvectors (pointing to the columns of $\begin{pmatrix} 3 & 1 \\ 2 & -1 \end{pmatrix}$)

All diagonalizable matrices can be viewed this way.

Thm: A is diagonalizable iff there is a basis for $\mathbb{R}^n$ consisting of eigenvectors.

Pf:

$$\begin{pmatrix} \lambda_1 & & 0 \\ & \ddots & \\ 0 & & \lambda_n \end{pmatrix} = \begin{pmatrix} | & & | \\ \vec{v}_1 & \cdots & \vec{v}_n \\ | & & | \end{pmatrix}^{-1} \begin{pmatrix} & & \\ & A & \\ & & \end{pmatrix} \begin{pmatrix} | & & | \\ \vec{v}_1 & \cdots & \vec{v}_n \\ | & & | \end{pmatrix}$$

eigenvalues (the diagonal entries $\lambda_1, \dots, \lambda_n$)

eigenvectors (the columns $\vec{v}_1, \dots, \vec{v}_n$)

change of basis matrix

$$\Updownarrow$$

$$\begin{pmatrix} | & & | \\ \vec{v}_1 & \cdots & \vec{v}_n \\ | & & | \end{pmatrix} \begin{pmatrix} \lambda_1 & & 0 \\ & \ddots & \\ 0 & & \lambda_n \end{pmatrix} = \begin{pmatrix} & & \\ & A & \\ & & \end{pmatrix} \begin{pmatrix} | & & | \\ \vec{v}_1 & \cdots & \vec{v}_n \\ | & & | \end{pmatrix}$$

$$\Updownarrow$$

$$\begin{pmatrix} | & & | \\ \lambda_1\vec{v}_1 & \cdots & \lambda_n\vec{v}_n \\ | & & | \end{pmatrix} = \begin{pmatrix} & & \\ & A & \\ & & \end{pmatrix} \begin{pmatrix} | & & | \\ \vec{v}_1 & \cdots & \vec{v}_n \\ | & & | \end{pmatrix}$$

Alt:) $A = [T]_{\mathcal{E}}^{\mathcal{E}}$, and if A is diagonalizable

we can also write

$$D = [T]_{\mathcal{V}}^{\mathcal{V}} = \begin{pmatrix} \lambda_1 & & 0 \\ & \ddots & \\ 0 & & \lambda_n \end{pmatrix}$$

for some basis $\mathcal{V} = \{\vec{v}_1, \ldots, \vec{v}_n\}$. D has

evident eigenvalues and eigenvectors

$$D \begin{pmatrix} 0 \\ \vdots \\ 1 \\ \vdots \\ 0 \end{pmatrix} = \lambda_i \begin{pmatrix} 0 \\ \vdots \\ 1 \\ \vdots \\ 0 \end{pmatrix}$$

and this can be rewritten as

$$[T]_{\mathcal{V}}^{\mathcal{V}} [\vec{v}_i]_{\mathcal{V}} = \lambda_i [\vec{v}_i]_{\mathcal{V}}$$

$$T(\vec{v}_i) = \lambda_i \vec{v}_i$$

$$[T]_{\mathcal{E}}^{\mathcal{E}} [\vec{v}_i]_{\mathcal{E}} = \lambda_i [\vec{v}_i]_{\mathcal{E}}$$

$$A \vec{v}_i = \lambda_i \vec{v}_i$$

So $\mathcal{V}$ is a basis of eigenvectors.

(And this argument can be reversed.)

So, for a diagonalizable matrix,

$$D = P^{-1} A P \qquad \text{and} \qquad A = P D P^{-1}$$

are related by the matrix of the basis of eigenvectors

$$P = \begin{pmatrix} | & & | \\ \vec{v}_1 & \cdots & \vec{v}_n \\ | & & | \end{pmatrix} = [I]^{\mathcal{S}}_{\mathcal{V}}$$

How do you remember where the "-1" goes?

Think of in terms of bases!

Ex:) Where is the "-1"?

$$A = P D P \; ?$$

$$[T]^{\mathcal{S}}_{\mathcal{S}} = [I]\,[T]^{\mathcal{V}}_{\mathcal{V}}\,[I]$$

So

$$A = P D P^{-1}$$

To determine if we can find such a basis, we use the following.

Thm: If we take bases for all of the eigenspaces and form a single set of vectors, that set is independent.

(Proved in book)

The question then becomes simply whether there are a total of n vectors in this set...

Recalling

$$\begin{array}{ccccc} \dim(E_{\lambda_1}) & & & \dim(E_{\lambda_k}) & \\ \text{/\!\!\wedge} & & & \text{/\!\!\wedge} & \\ m_1 & + \cdots & + & m_k & = n \end{array}$$

we see that this will only happen if

$$\dim(E_{\lambda_i}) = m_i$$

for all eigenspaces.

Ex:) From our first example, the char. poly was

$$(\lambda+1)(\lambda-4)$$

So we had

$$\lambda_1 = -1 \qquad m_1 = 1$$

$$\lambda_2 = 4 \qquad m_2 = 1$$

Each of these eigenvalues had an eigenvector, so

$$\dim(E_{\lambda_1}) = 1 = m_1, \quad \dim(E_{\lambda_2}) = 1 = m_2$$

Ex:) Recall that for

$$A = \begin{pmatrix} 1 & 1 \\ 0 & 1 \end{pmatrix}$$

we found the char. poly was $(\lambda-1)^2$, so there is only one eigenvalue:

$$\lambda_1 = 1 \qquad m_1 = 2$$

But there was only 1 eigenvector... so

$$\dim(E_{\lambda_1}) = 1 \neq m_1$$

So A is <u>not</u> diagonalizable.

Applications

Suppose you want to raise a matrix to a high power.

Diagonalizability helps!

Ex i) $\begin{pmatrix} -1 & 0 \\ 0 & 4 \end{pmatrix} = \begin{pmatrix} 3 & 1 \\ 2 & -1 \end{pmatrix}^{-1} \begin{pmatrix} 1 & -3 \\ -2 & 2 \end{pmatrix} \begin{pmatrix} 3 & 1 \\ 2 & -1 \end{pmatrix}$

$$D = P^{-1} \quad A \quad P$$

So also $A = PDP^{-1}$

Then
$$\begin{aligned} A^2 &= (PDP^{-1})(PDP^{-1}) \\ &= PD(P^{-1}P)DP^{-1} \\ &= PD^2P^{-1} \end{aligned}$$

Similarly,
$$A^n = PD^nP^{-1}$$

So
$$\begin{pmatrix} 1 & -3 \\ -2 & 2 \end{pmatrix}^n = \begin{pmatrix} 3 & 1 \\ 2 & -1 \end{pmatrix} \begin{pmatrix} (-1)^n & 0 \\ 0 & 4^n \end{pmatrix} \begin{pmatrix} 3 & 1 \\ 2 & -1 \end{pmatrix}^{-1}$$

Ex: What is the nth Fibonacci number?

$0, 1, 1, 2, 3, 5, 8, \ldots$ $\qquad f_{n+1} = f_n + f_{n-1}$

$f_0, f_1, f_2, \ldots$

Note $\begin{pmatrix} f_{n+1} \\ f_n \end{pmatrix} = \begin{pmatrix} 1 & 1 \\ 1 & 0 \end{pmatrix} \begin{pmatrix} f_n \\ f_{n-1} \end{pmatrix}$

So $\begin{pmatrix} f_{n+1} \\ f_n \end{pmatrix} = \begin{pmatrix} 1 & 1 \\ 1 & 0 \end{pmatrix}^n \begin{pmatrix} 1 \\ 0 \end{pmatrix}$

and thus f_n is the bottom left entry of A^n, with

$$A = \begin{pmatrix} 1 & 1 \\ 1 & 0 \end{pmatrix}$$

$\lambda_1 = g$, $\vec{v}_1 = \begin{pmatrix} g \\ 1 \end{pmatrix}$; $\lambda_2 = -1/g$, $\vec{v}_2 = \begin{pmatrix} 1 \\ -g \end{pmatrix}$ $\qquad g = \frac{1+\sqrt{5}}{2}$

Then $A^n = \begin{pmatrix} g & 1 \\ 1 & -g \end{pmatrix} \begin{pmatrix} g^n & 0 \\ 0 & (-1/g)^n \end{pmatrix} \begin{pmatrix} g & 1 \\ 1 & -g \end{pmatrix}^{-1}$

For large n,

$$\cong \begin{pmatrix} g & 1 \\ 1 & -g \end{pmatrix} \begin{pmatrix} g^n & 0 \\ 0 & 0 \end{pmatrix} \begin{pmatrix} g & 1 \\ 1 & -g \end{pmatrix} \Big/ (g^2+1)$$

$$\cong \frac{g^n}{g^2+1} \begin{pmatrix} g^2 & g \\ g & 1 \end{pmatrix}$$

So $f_n \cong \frac{g^{n+1}}{g^2+1}$.

Ex i) Suppose two quantities a and b take values in stages, and each value depends on the values in the previous stage.

$$\begin{pmatrix} a_{k+1} \\ b_{k+1} \end{pmatrix} = \begin{pmatrix} & M & \end{pmatrix} \begin{pmatrix} a_k \\ b_k \end{pmatrix}$$

Then

$$\begin{pmatrix} a_k \\ b_k \end{pmatrix} = M^k \begin{pmatrix} a_0 \\ b_0 \end{pmatrix}$$

Again, diagonalization allows us to compute this.

$$M = PDP^{-1}$$

$$M^k = PD^kP^{-1}$$

$$\begin{pmatrix} a_k \\ b_k \end{pmatrix} = PD^kP^{-1} \begin{pmatrix} a_0 \\ b_0 \end{pmatrix}$$

Jordan Canonical Form

What if A is not diagonalizable?

Can <u>always</u> be put (by similarity / change of basis) into the (more general) Jordan form.

Like "diagonal form", there are rules defining what "Jordan form" is.

Eigenvalue block structure

Say A has $p(\lambda) = (\lambda - r_1)^{m_1} (\lambda - r_2)^{m_2} \cdots (\lambda - r_k)^{m_k}$

The Jordan form has blocks "on the diagonal" that "correspond" to these eigenvalues.

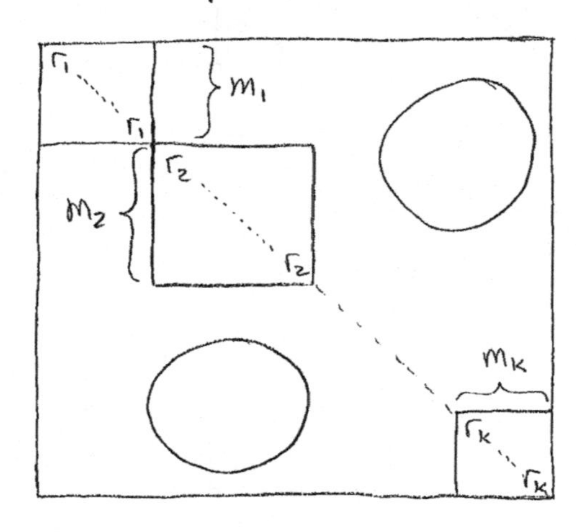

- diagonal entries are eigenvalues.
- block sizes are multiplicities.
- all zeroes outside of these eigenvalue blocks.

What else is inside of an eigenvalue block?

Basic Jordan block structure

Inside of an eigenvalue block, there are basic Jordan blocks on the diagonal.

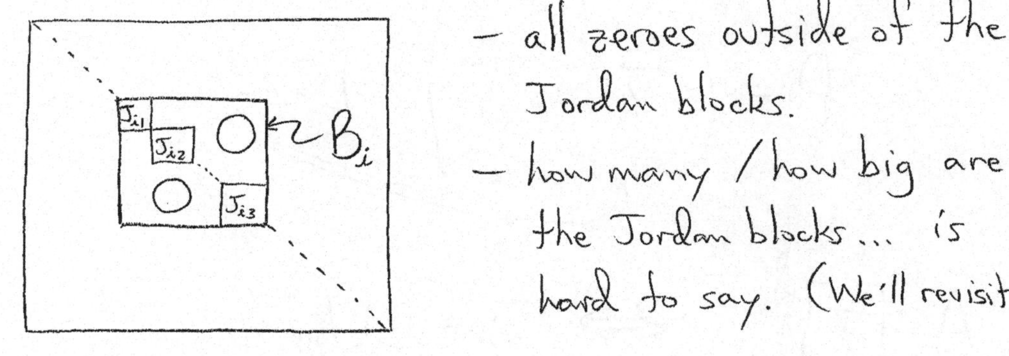

- all zeroes outside of the Jordan blocks.
- how many / how big are the Jordan blocks ... is hard to say. (We'll revisit soon!)

What is inside of a basic Jordan block?

They all have the same structure!

$$J = \begin{pmatrix} r_i & 1 & & O \\ & r_i & 1 & \\ & & \ddots & \ddots \\ & & & 1 \\ O & & & r_i \end{pmatrix}$$

- O ← zeroes everywhere except for...
- 1 ← 1's just above diag...
- r_i ← eigenvalue on diag.

Ex: The following are all in Jordan form.

$$\begin{pmatrix} 3&1&0&0&0 \\ 0&3&1&0&0 \\ 0&0&3&0&0 \\ 0&0&0&5&0 \\ 0&0&0&0&5 \end{pmatrix}, \quad \begin{pmatrix} 3&1&0&0&0 \\ 0&3&0&0&0 \\ 0&0&3&0&0 \\ 0&0&0&5&1 \\ 0&0&0&0&5 \end{pmatrix}, \quad \begin{pmatrix} 5&0&0&0&0 \\ 0&5&0&0&0 \\ 0&0&3&0&0 \\ 0&0&0&3&1 \\ 0&0&0&0&3 \end{pmatrix}$$

So, for all matrices A,

$$J = P^{-1} A P \quad \text{and} \quad A = P J P^{-1}$$

are related by the matrix of the Jordan basis

$$P = \begin{pmatrix} | & & | \\ \vec{v}_1 & \cdots & \vec{v}_n \\ | & & | \end{pmatrix} = [I]_{\mathcal{V}}^{\mathcal{S}}$$

and

$$A = [T]_{\mathcal{S}}^{\mathcal{S}} \iff J = [T]_{\mathcal{V}}^{\mathcal{V}}$$

How do you remember where the "-1" goes?
Think of in terms of bases!

Ex: Where is the "-1"?

$$A = P J P \quad ?$$

$$[T]_{\mathcal{S}}^{\mathcal{S}} = [I] \, [T]_{\mathcal{V}}^{\mathcal{V}} \, [I]$$

So

$$A = P J P^{-1}$$

Rearrangements

You can "rearrange" blocks by a change of basis that comes from "reordering" the basis vectors.

Ex:) Say a matrix A is put into Jordan form J by $\mathcal{V} = \{\vec{v}_1, \vec{v}_2, \vec{v}_3, \vec{v}_4, \vec{v}_5\}$. That is,

$$A = [T]_{\mathcal{S}}^{\mathcal{S}} \qquad J_{\mathcal{V}} = [T]_{\mathcal{V}}^{\mathcal{V}} = \begin{pmatrix} 2 & 1 & & & \\ & 2 & & & \\ & & 2 & & \\ & & & 6 & 1 \\ & & & & 6 \end{pmatrix}$$

Reordering $\mathcal{V}$ to $\mathcal{W} = \{\vec{v}_4, \vec{v}_5, \vec{v}_3, \vec{v}_1, \vec{v}_2\}$ results in

$$J_{\mathcal{W}} = [T]_{\mathcal{W}}^{\mathcal{W}} = \begin{pmatrix} 6 & 1 & & & \\ & 6 & & & \\ & & 2 & & \\ & & & 2 & 1 \\ & & & & 2 \end{pmatrix}$$

Note we have rearranged both the eigenvalue blocks and the Jordan blocks inside them.

You can confirm the assertion above by either

- finding $[I]_{\mathcal{V}}^{\mathcal{W}}$ and using change of basis

or

- interpreting $[T]_{\mathcal{V}}^{\mathcal{V}}$ as a list of equations, and then rewriting appropriate to $\mathcal{W}$.

Try both! (See also AHP's.)

Jordan form is used to represent equivalence classes (by similarity / change of basis) of matrices.

Since these "rearranged" forms are similar and thus represent all the same matrices, we consider them equivalent.

<u>Thm:</u>) Up to rearrangements, every matrix can be put into <u>exactly</u> one Jordan form.

(The proof of this is beyond the scope of this course.)

<u>Ex:</u>) The Jordan forms

$$\begin{pmatrix} 3 & 1 & 0 \\ 0 & 3 & 0 \\ 0 & 0 & 3 \end{pmatrix} \quad \text{and} \quad \begin{pmatrix} 3 & 1 & 0 \\ 0 & 3 & 1 \\ 0 & 0 & 3 \end{pmatrix}$$

are <u>not</u> similar, because their blocks are not rearrangements of each other.

Relationship to eigenvectors

Say $J = [T]_{\mathcal{V}}^{\mathcal{V}}$ is in Jordan form; we refer to the basis $\mathcal{V}$ as a Jordan basis.

$$\mathcal{V} = \{\vec{v}_1, \dots, \vec{v}_n\}$$

Recall that

$$[T]_{\mathcal{V}}^{\mathcal{V}} = \begin{pmatrix} & \textcircled{i} & \\ & | & \\ & | & \end{pmatrix}$$

the ith column of the matrix gives $T(\vec{v}_i)$ in terms of $\mathcal{V}$; that is the column gives coefficients in a l.c. of $\mathcal{V}$ that equals $T(\vec{v}_i)$.

So, for example: $T(\vec{v}_i) = \lambda \vec{v}_i$; $T(\vec{v}_j) = 1\vec{v}_{j-1} + \lambda \vec{v}_j$

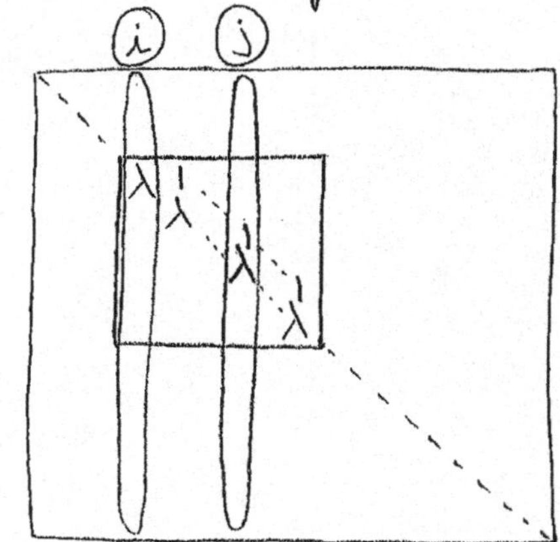

So
- $\vec{v}_i$ is an eigenvector
- $\vec{v}_j$ is not.

Given the locations of the 1's in Jordan blocks, the only Jordan basis vectors that can be eigenvectors are those corresponding to the start of a Jordan block.

In fact,

Thm: The number of eigenvectors with a given eigenvalue λ equals the number of basic Jordan blocks in the λ eigenvalue block.

Ex: Suppose A is equivalent to

$$\begin{pmatrix} 4 & 1 & & \\ & 4 & & \\ & & 4 & \\ & & & 3 \end{pmatrix}$$

Then – the eigenvalue 4 has 2 eigenvectors
– the eigenvalue 3 has 1 eigenvector.

Ex: What does it mean about a matrix A if the Jordan form is

$$J = \begin{pmatrix} 4 & 1 & 0 \\ 0 & 4 & 0 \\ 0 & 0 & 4 \end{pmatrix}$$

A is the matrix for a l.t. T w.r.t. the standard basis $\mathcal{S}$; J is a change of basis on A, using the Jordan basis $\mathcal{V} = \{\vec{v}_1, \vec{v}_2, \vec{v}_3\}$. That is,

$$J = [T]_{\mathcal{V}}^{\mathcal{V}}$$

So (writing in terms of the standard basis),

$$A\vec{v}_1 = 4\vec{v}_1$$
$$A\vec{v}_2 = 1\vec{v}_1 + 4\vec{v}_2$$
$$A\vec{v}_3 = 4\vec{v}_3$$

Notice — $\vec{v}_1, \vec{v}_3$ are eigenvectors, as we see from the fact that the corresponding columns have nonzero entries only on the diagonal.

We will <u>not</u> prove the theorem about Jordan canonical form.

We will <u>not</u> discuss how to find the basis that puts a non-diagonalizable matrix into Jordan form.

Still, in some cases, we can still do this.

Ex: Note that if A is diagonalizable, then its diagonal form <u>is</u> Jordan form.
(Every basic Jordan block is 1×1, and every vector in the Jordan basis is an eigenvector.)

Ex: Suppose A has characteristic polynomial

$$p(\lambda) = (\lambda-1)^2(\lambda-3)$$

We have $\lambda_1 = 1$, $m_1 = 2$
$\lambda_2 = 3$, $m_2 = 1$

So there are only 2 possible Jordan forms:

$$\begin{pmatrix} 1 & 0 & 0 \\ 0 & 1 & 0 \\ 0 & 0 & 3 \end{pmatrix}, \begin{pmatrix} 1 & 1 & 0 \\ 0 & 1 & 0 \\ 0 & 0 & 3 \end{pmatrix}$$

If λ_1 has 2 eigenvectors, then it is the former.
If λ_1 has 1 eigenvector, then it is the latter.

<u>Obs.</u> Within an eigenvalue block B_i with multiplicity m_i, the number of possibilities is the <u>partition function</u> (the number of different ways positive integers can add up to m_i.)

<u>Ex:</u>) How many Jordan forms are possible if the char. poly. is

$$(\lambda-7)^4(\lambda-5)^3 \quad ?$$

$$\begin{aligned} 4 &= 4 \\ &= 3+1 \\ &= 2+2 \\ &= 2+1+1 \\ &= 1+1+1+1 \end{aligned} \qquad \begin{aligned} 3 &= 3 \\ &= 2+1 \\ &= 1+1+1 \end{aligned}$$

There are 5 possible B_1 ; 3 possible B_2.
So there are 15 possible Jordan forms.

9.1 – Inner Product Spaces

Recall the following facts about the dot product in $\mathbb{R}^n$:

Def: $\vec{u}\cdot\vec{v} = u_1v_1 + \cdots + u_nv_n$

Facts: $\vec{u}\cdot\vec{v} = \vec{v}\cdot u$

$(\vec{u}+\vec{v})\cdot\vec{w} = \vec{u}\cdot\vec{w} + \vec{v}\cdot\vec{w}$

$(c\vec{u})\cdot\vec{v} = c(\vec{u}\cdot\vec{v})$ (note difference in book!)

$\vec{v}\cdot\vec{v} \geq 0$; equality iff $\vec{v}=\vec{0}$

More Facts: $\vec{0}\cdot\vec{v} = 0$

$\vec{v}\cdot(\vec{u}+\vec{w}) = \vec{v}\cdot\vec{u} + \vec{v}\cdot\vec{w}$

$\vec{v}\cdot(c\vec{w}) = c(\vec{v}\cdot\vec{w})$

Still More Facts: $\|\vec{v}\| = \sqrt{\vec{v}\cdot\vec{v}} = \sqrt{\vec{v}^T\vec{v}}$

$|\vec{v}\cdot\vec{w}| \leq \|\vec{v}\|\,\|\vec{w}\|$

$\vec{v}\cdot\vec{w} = \|\vec{v}\|\,\|\vec{w}\|\cos\theta$

$\theta = \arccos\left(\frac{\vec{v}\cdot\vec{w}}{\|\vec{v}\|\,\|\vec{w}\|}\right)$

We will use the first 4 facts above to define a generalization of this dot product for other kinds of vector spaces.

Def: V a vector space, suppose we have a function $\langle , \rangle$ on pairs of vectors, with real values, satisfying

1. $\langle u,v \rangle = \langle v,u \rangle$
2. $\langle u+v, w \rangle = \langle u,w \rangle + \langle v,w \rangle$
3. $\langle cu, v \rangle = c \langle u,v \rangle$
4. $\langle v,v \rangle \geq 0$; equality iff $v = 0$ (different in book!)

The function $\langle , \rangle$ is called an <u>inner product</u>, and V is called an <u>inner product space.</u>

Ex: $\mathbb{R}^n$, $\langle \vec{u}, \vec{v} \rangle = \vec{u} \cdot \vec{v}$ is an inner product space

Ex: On $C[a,b]$, define

$$\langle f,g \rangle = \int_a^b f(x)g(x)\,dx$$

Can check this is an inner product. (Called the <u>L^2 inner product</u>.)

Ex: Note dot product is <u>not</u> preserved by change of basis...

That is, $[\vec{v}]_{\mathcal{B}} \cdot [\vec{w}]_{\mathcal{B}} \neq [\vec{v}]_{\mathcal{U}} \cdot [\vec{w}]_{\mathcal{U}}$

But this <u>does</u> give us a different inner product;

$$\langle \vec{v}, \vec{w} \rangle = [\vec{v}]_{\mathcal{U}} \cdot [\vec{w}]_{\mathcal{U}}$$

(Think about this in Exercise 9.)

Note, the "More Facts" about dot products can be shown for all inner products, using the definition.

Note also that inner products are linear in each entry:

$$\langle c_1\vec{v}_1 + \cdots + c_n\vec{v}_n , \vec{w} \rangle = c_1\langle \vec{v}_1, \vec{w}\rangle + \cdots + c_n\langle \vec{v}_n, \vec{w}\rangle$$

$$\langle \vec{w}, c_1\vec{v}_1 + \cdots + c_n\vec{v}_n \rangle = c_1\langle \vec{w}, \vec{v}_1\rangle + \cdots + c_n\langle \vec{w}, \vec{v}_n\rangle$$

We have defined inner products for vector spaces that might not be Euclidean, and so might not have pre-existing notions of length and angle. But we can use the "Still More Facts" to <u>motivate</u> definitions for these.

Def: For an inner product space V, define

$$\|\vec{v}\| = \sqrt{\langle \vec{v}, \vec{v}\rangle}$$

We call this the <u>norm</u> or <u>magnitude</u> of $\vec{v}$.
(Can think of this as a sort of "length".)

Ex: $C^0[a,b]$ with the L^2 inner product gives us

$$\|f\| = \left(\int_a^b (f(x))^2\,dx\right)^{1/2}$$

This is called the <u>L^2 norm</u> (there are similar norms called L^p norms that do not come from inner products.)

Def) The <u>angle</u> between two vectors $\vec{v}, \vec{w}$ is defined as

$$\theta = \arccos\left(\frac{\langle \vec{v}, \vec{w} \rangle}{\|\vec{v}\| \, \|\vec{w}\|}\right)$$

Ex) What is the angle between $\sin(x)$, $\sin(x+\phi)$ in $C^0[0, 2\pi]$?

① $\|\sin x\| = \left(\int_0^{2\pi} \sin^2 x \, dx\right)^{1/2} = \sqrt{\pi}$

② $\|\sin(x+\phi)\| = \left(\int_0^{2\pi} \sin^2(x+\phi) \, dx\right)^{1/2} = \sqrt{\pi}$

③ $\langle \sin x, \sin(x+\phi) \rangle = \int_0^{2\pi} (\sin x)(\sin x \cos\phi + \cos x \sin\phi) \, dx$

$$= (\cos\phi)\left(\int_0^{2\pi} \sin^2 x \, dx\right) + (\sin\phi)\left(\int_0^{2\pi} \sin x \cos x \, dx\right)$$

$$= (\cos\phi)(\pi)$$

④ $\theta = \arccos\left(\frac{\pi \cos\phi}{\sqrt{\pi}\sqrt{\pi}}\right) = \phi$ ☺

How do we know the above arccos will always be defined?

The Cauchy-Schwarz inequality gives us this.

Thm) For <u>any</u> inner product space, we have

$$|\langle u, v \rangle| \leq \|u\| \, \|v\|$$

(proved in book)

Another example of angle involves statistics.

Ex:) Suppose X and Y are real random variables on the sample space S. The expression

$$\langle X, Y \rangle = E\big((X-\mu_X)(Y-\mu_Y) \big)$$

is an inner product on these random variables (called the covariance). Other statistical notions derive from this

$$\langle X, Y \rangle = \text{covariance} = \text{cov}(X, Y)$$

$$\langle X, X \rangle = \text{variance}$$

$$\|X\| = \sqrt{\langle X, X \rangle} = \text{standard deviation}$$

$$\cos\theta = \frac{\langle X, Y \rangle}{\|X\| \, \|Y\|} = \text{correlation}$$

This connection to angle leads naturally to a surprising fact about random variables — when X, Y are positively correlated, and Y, Z are positively correlated, X, Z might still be negatively correlated!

From a vector point of view, positive correlation relates to an acute angle. Geometrically then:

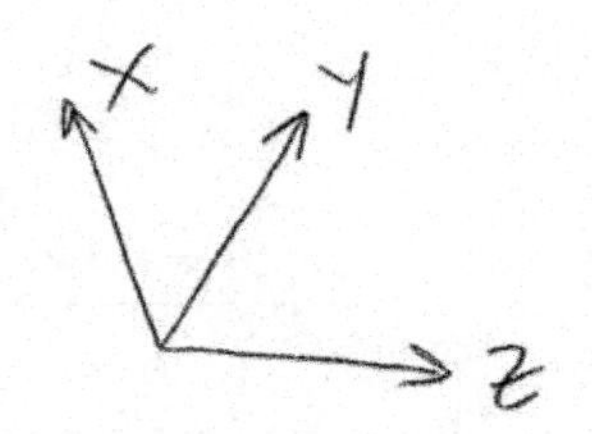

An example: Let $S = \{ \vec{v} \in \mathbb{R}^2 \mid \|\vec{v}\| \leq 1 \}$, and $\vec{x}, \vec{y}, \vec{z}$ three vectors in $\mathbb{R}^2$ arranged as above.

Then define
$$X = \vec{x} \cdot \vec{v}$$
$$Y = \vec{y} \cdot \vec{v}$$
$$Z = \vec{z} \cdot \vec{v}$$

These are random variables on S, with

$$\operatorname{cov}(X, Y) > 0$$
and
$$\operatorname{cov}(Y, Z) > 0$$

but
$$\operatorname{cov}(X, Z) < 0$$

9.2 — Orthonormal Bases

Recall $\vec{v}, \vec{w} \in V$ are <u>orthogonal</u> if $\langle \vec{v}, \vec{w} \rangle = 0$.

Similarly:

<u>Def:</u> $\mathcal{V} = \{\vec{v}_1, \dots, \vec{v}_n\}$ is an <u>orthogonal basis</u> for V iff it is a basis and the vectors are orthogonal.

<u>Def:</u> $\mathcal{V} = \{\vec{v}_1, \dots, \vec{v}_n\}$ is an <u>orthonormal basis</u> for V iff it is orthogonal <u>and</u> $\|\vec{v}_i\| = 1$.

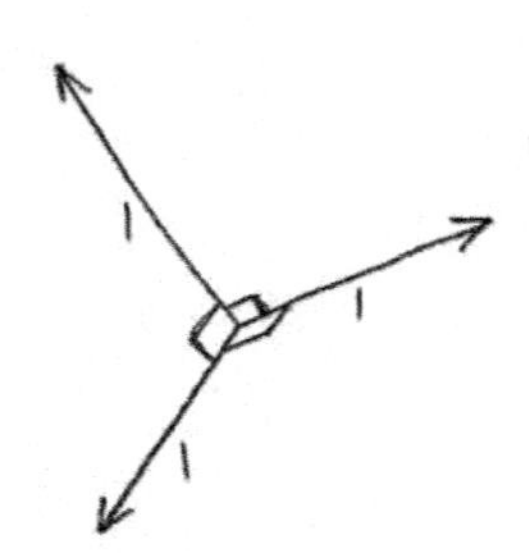

<u>Ex:</u> $\mathcal{E}$ is an orthonormal basis for $\mathbb{R}^n$

<u>Ex:</u> $\left\{ \begin{pmatrix} 1/\sqrt{2} \\ 1/\sqrt{2} \\ 0 \end{pmatrix}, \begin{pmatrix} 1/\sqrt{6} \\ -1/\sqrt{6} \\ 2/\sqrt{6} \end{pmatrix}, \begin{pmatrix} 1/\sqrt{3} \\ -1/\sqrt{3} \\ -1/\sqrt{3} \end{pmatrix} \right\}$ is an orthonormal basis for $\mathbb{R}^3$

Here are some properties of orthonormal bases.

Thm:) If $\mathcal{V} = \{\vec{v}_1, \ldots, \vec{v}_n\}$ is an orthonormal basis for V, then for any $\vec{v} \in V$, the coords. are the projections.

That is

$$[\vec{v}]_{\mathcal{V}} = \begin{pmatrix} \langle \vec{v}, \vec{v}_1 \rangle \\ \vdots \\ \langle \vec{v}, \vec{v}_n \rangle \end{pmatrix}$$

Pf:) Say $\vec{v} = c_1\vec{v}_1 + \cdots + c_n\vec{v}_n$. Then

$$\langle \vec{v}, \vec{v}_i \rangle = \langle c_1\vec{v}_1 + \cdots + c_n\vec{v}_n \, , \, \vec{v}_i \rangle$$

$$= c_1 \underbrace{\langle \vec{v}_1, \vec{v}_i \rangle}_{0} + \cdots + c_i \underbrace{\langle \vec{v}_i, \vec{v}_i \rangle}_{1} + \cdots + c_n \underbrace{\langle \vec{v}_n, \vec{v}_i \rangle}_{0}$$

$$= c_i$$

Ex:) How do we write $\vec{v} = (1,2,3)$ as a l.c. of the orth. basis on prev. page?

$$\left.\begin{aligned} c_1 &= \begin{pmatrix} 1 \\ 2 \\ 3 \end{pmatrix} \cdot \begin{pmatrix} 1/\sqrt{2} \\ 1/\sqrt{2} \\ 0 \end{pmatrix} = \frac{3}{\sqrt{2}} \\ c_2 &= \begin{pmatrix} 1 \\ 2 \\ 3 \end{pmatrix} \cdot \begin{pmatrix} 1/\sqrt{6} \\ -1/\sqrt{6} \\ 2/\sqrt{6} \end{pmatrix} = \frac{5}{\sqrt{6}} \\ c_3 &= \begin{pmatrix} 1 \\ 2 \\ 3 \end{pmatrix} \cdot \begin{pmatrix} 1/\sqrt{3} \\ -1/\sqrt{3} \\ -1/\sqrt{3} \end{pmatrix} = \frac{-4}{\sqrt{3}} \end{aligned}\right\} \Rightarrow [\vec{v}]_{\mathcal{V}} = \begin{pmatrix} 3/\sqrt{2} \\ 5/\sqrt{6} \\ -4/\sqrt{3} \end{pmatrix}$$

Geometrically:

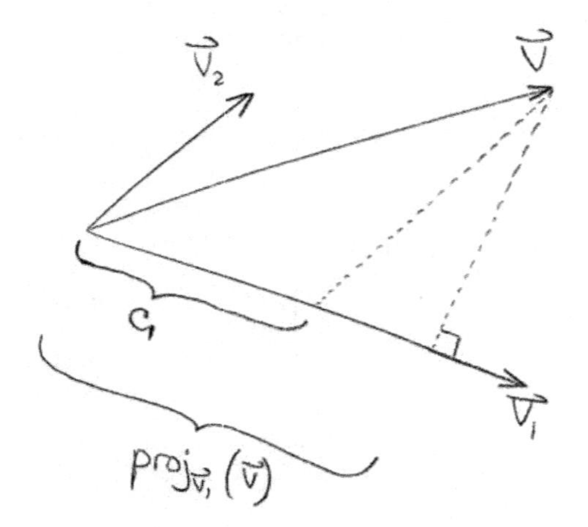

When $\mathcal{V} = \{\vec{v}_1, \vec{v}_2\}$ is <u>not</u> orthonormal, the coordinates are <u>not</u> necessarily equal to the projections.

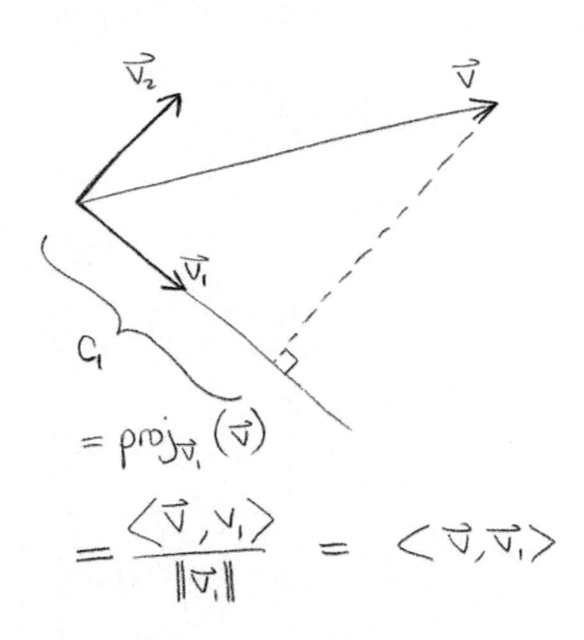

$$= \text{proj}_{\vec{v}_1}(\vec{v})$$

$$= \frac{\langle \vec{v}, v_1 \rangle}{\|\vec{v}_1\|} = \langle \vec{v}, \vec{v}_1 \rangle$$

Def: A matrix A is an <u>orthogonal matrix</u> if the column vectors form an orthonormal basis.

$$A = \begin{pmatrix} | & & | \\ \vec{v}_1 & \cdots & \vec{v}_n \\ | & & | \end{pmatrix}, \quad \{\vec{v}_1, \ldots, \vec{v}_n\} \text{ is an orthonormal basis}$$

Thm: A is orthogonal $\iff$ $A^T A = I$

(NB, the book reverses this def and thm)

Pf:

$$A^T A = \begin{pmatrix} \text{---} & \vec{v}_1 & \text{---} \\ & \vdots & \\ \text{---} & \vec{v}_n & \text{---} \end{pmatrix} \begin{pmatrix} | & & | \\ \vec{v}_1 & \cdots & \vec{v}_n \\ | & & | \end{pmatrix}$$

The entries in this product are rows dot columns, which are

$$\vec{v}_i \cdot \vec{v}_j$$

which are 0 if $i \neq j$ and 1 if $i = j$.

So $A^T A = I$.

A nice application of this involves dot products.

Thm: If A is orthogonal, then it preserves dot products.

Pf:

$$\begin{aligned}(A\vec{v}) \cdot (A\vec{w}) &= (A\vec{v})^T (A\vec{w}) \\ &= \vec{v}^T A^T A \vec{w} \\ &= \vec{v}^T \vec{w} \\ &= \vec{v} \cdot \vec{w}\end{aligned}$$

Note, since lengths and angles are written in terms of dot products, these are preserved too! Such matrices are rigid motions on $\mathbb{R}^n$.

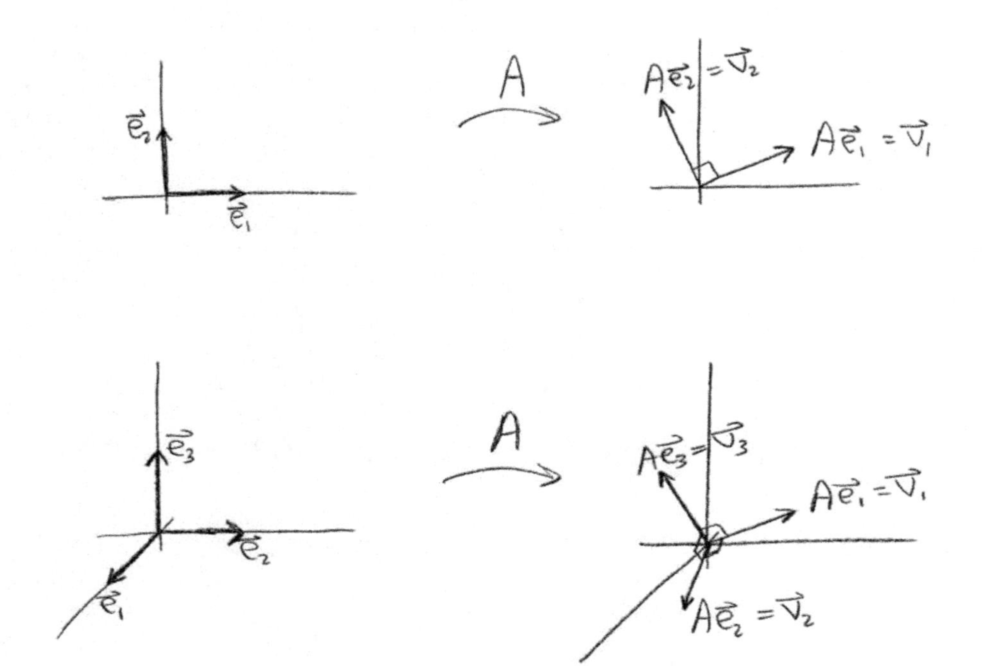

They are <u>all</u> rotations with a possible reflection.

("SO_3" is the set of all rotations (no reflection) in $\mathbb{R}^3$. There is a proof of a special property of SO_3 that <u>literally</u> involves handwaving... (demo in class).)

How can we convert a basis that is <u>not</u> orthonormal into another one that is?

Gram-Schmidt Orthonormalization

Geometrically:

Given $\vec{w}_2$, $\vec{w}_1$

$\vec{w}_2$, $\vec{v}_1$, 1

$\vec{w}_2$, $\vec{v}_1$

$\vec{v}_2$, $\vec{v}_1$ ← orthonormal!

Basically, we use $\vec{v}_1$ to adjust $\vec{w}_2$ to something orthogonal to $\vec{v}_1$, and then divide by length to "normalize".

Algebraically:

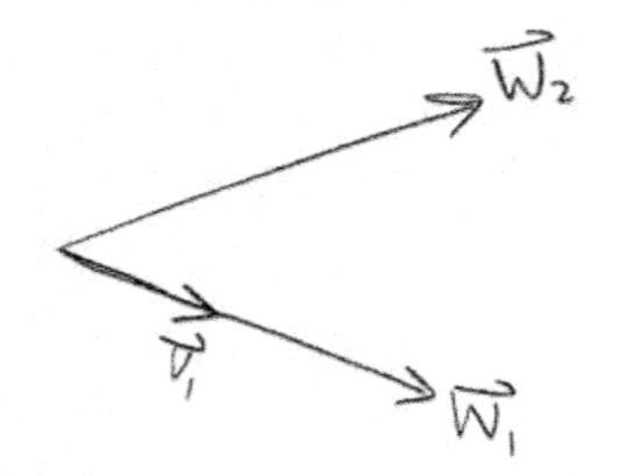

$$\vec{v}_1 = \frac{\vec{w}_1}{\|\vec{w}_1\|}$$

$$\vec{w}_2 - \left(\langle \vec{w}_2, \vec{v}_1 \rangle \vec{v}_1\right) = \vec{x}_2$$

$\langle \vec{w}_2, \vec{v}_1 \rangle \vec{v}_1$

$\vec{w}_2$

$\vec{v}_1$

$$\vec{v}_2 = \frac{\vec{x}_2}{\|\vec{x}_2\|}$$

What if we have a basis with 3 vectors?

Same process for 1st two... Then:

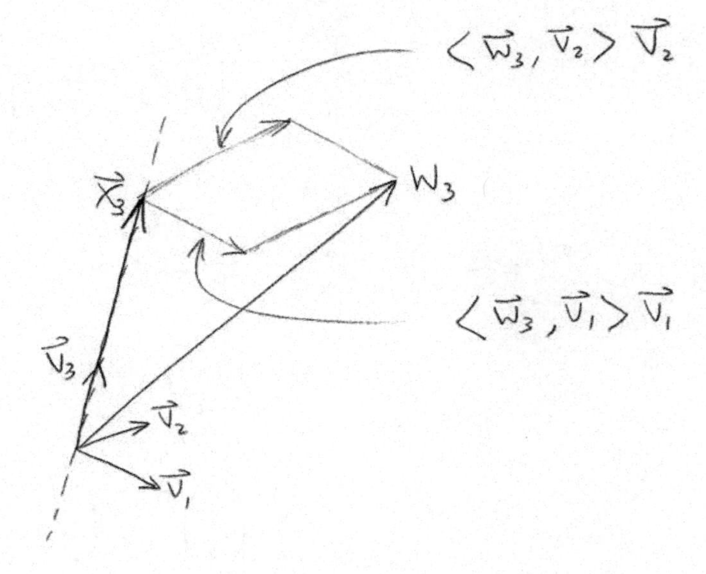

$$\vec{x}_3 = \vec{w}_3 - \langle \vec{w}_3, \vec{v}_1 \rangle \vec{v}_1 - \langle \vec{w}_3, \vec{v}_2 \rangle \vec{v}_2$$

$$\vec{v}_3 = \frac{\vec{x}_3}{\|\vec{x}_3\|}$$

Similarly for bases with even more vectors...

Ex:) Apply the Gram-Schmidt process to the basis

$$\mathcal{W} = \left\{ \begin{pmatrix} 2 \\ 3 \\ 6 \end{pmatrix}, \begin{pmatrix} 1 \\ 1 \\ 2 \end{pmatrix}, \begin{pmatrix} 1 \\ 0 \\ 4 \end{pmatrix} \right\} = \{\vec{w}_1, \vec{w}_2, \vec{w}_3\}$$

Compute $\vec{v}_1$:
$$\vec{v}_1 = \frac{\vec{w}_1}{\|\vec{w}_1\|} = \begin{pmatrix} 2 \\ 3 \\ 6 \end{pmatrix} / 7 = \begin{pmatrix} 2/7 \\ 3/7 \\ 6/7 \end{pmatrix}$$

Compute $\vec{v}_2$: First,
$$\vec{x}_2 = \vec{w}_2 - \langle \vec{w}_2, \vec{v}_1 \rangle \vec{v}_1$$
$$= \begin{pmatrix} 1 \\ 1 \\ 2 \end{pmatrix} - \left(\begin{pmatrix} 1 \\ 1 \\ 2 \end{pmatrix} \cdot \begin{pmatrix} 2/7 \\ 3/7 \\ 6/7 \end{pmatrix} \right) \begin{pmatrix} 2/7 \\ 3/7 \\ 6/7 \end{pmatrix}$$
$$= \begin{pmatrix} 1 \\ 1 \\ 2 \end{pmatrix} - \left(\frac{17}{7} \right) \begin{pmatrix} 2/7 \\ 3/7 \\ 6/7 \end{pmatrix}$$
$$= \begin{pmatrix} 1 \\ 1 \\ 2 \end{pmatrix} - \begin{pmatrix} 34/49 \\ 51/49 \\ 102/49 \end{pmatrix} = \begin{pmatrix} 15/49 \\ -2/49 \\ -4/49 \end{pmatrix}$$

Then
$$\vec{v}_2 = \frac{\vec{x}_2}{\|\vec{x}_2\|} = \frac{\vec{x}_2}{\sqrt{245}/49} = \begin{pmatrix} 15/7\sqrt{5} \\ -2/7\sqrt{5} \\ -4/7\sqrt{5} \end{pmatrix}$$

Compute $\vec{v}_3$:

$$\vec{x}_3 = \vec{w}_3 - \langle \vec{w}_3, \vec{v}_1 \rangle \vec{v}_1 - \langle \vec{w}_3, \vec{v}_2 \rangle \vec{v}_2$$

$$= \begin{pmatrix} 1 \\ 0 \\ 4 \end{pmatrix} - \left(\begin{pmatrix} 1 \\ 0 \\ 4 \end{pmatrix} \cdot \begin{pmatrix} 2 \\ 3 \\ 6 \end{pmatrix} / 7 \right) \begin{pmatrix} 2 \\ 3 \\ 6 \end{pmatrix} / 7$$

$$- \left(\begin{pmatrix} 1 \\ 0 \\ 4 \end{pmatrix} \cdot \begin{pmatrix} 15 \\ -2 \\ -4 \end{pmatrix} / 7\sqrt{5} \right) \begin{pmatrix} 15 \\ -2 \\ -4 \end{pmatrix} / 7\sqrt{5}$$

$$= \begin{pmatrix} 1 \\ 0 \\ 4 \end{pmatrix} - \frac{26}{49} \begin{pmatrix} 2 \\ 3 \\ 6 \end{pmatrix} - \frac{-1}{245} \begin{pmatrix} 15 \\ -2 \\ -4 \end{pmatrix}$$

$$= \begin{pmatrix} 1 \\ 0 \\ 4 \end{pmatrix} - \begin{pmatrix} 260 \\ 390 \\ 780 \end{pmatrix} / 245 + \begin{pmatrix} 15 \\ -2 \\ -4 \end{pmatrix} / 245$$

$$= \begin{pmatrix} 0 \\ -392 \\ 196 \end{pmatrix} / 245$$

$$= \begin{pmatrix} 0 \\ -8 \\ 4 \end{pmatrix} / 5 = \begin{pmatrix} 0 \\ -8/5 \\ 4/5 \end{pmatrix}$$

$$\vec{v}_3 = \frac{\vec{x}_3}{\|\vec{x}_3\|} = \frac{\vec{x}_3}{\sqrt{80}/5} = \frac{\vec{x}_3}{4/\sqrt{5}} = \begin{pmatrix} 0 \\ -2 \\ 1 \end{pmatrix} / \sqrt{5}$$

9.3 — Complex Inner Product Spaces, Symmetric Matrices

Recall that the vector spaces we have talked about up to now were "vector spaces over $\mathbb{R}$"; that is, the scalar multiplication was only by real numbers.

A similar (but different) thing is a "vector space over $\mathbb{C}$", in which we allow scalar multiplication by complex numbers.

Ex:) $\mathbb{C}$ itself is a vector space over $\mathbb{C}$.

It "has 1 complex dimension" ($\{1\}$ is a basis).

On the other hand, we could have viewed $\mathbb{C}$ as a vector space over $\mathbb{R}$; as such it "has 2 real dimensions" ($\{1, i\}$ is a basis).

Ex:) $\mathbb{C}^n = \left\{ (z_1, \ldots, z_n) \mid z_i \in \mathbb{C} \right\}$ is a vector space over $\mathbb{C}$ of dimension n.

It is a vector space over $\mathbb{R}$ of dimension $2n$.

We have to treat vector spaces over $\mathbb{C}$ totally differently.

Specifically, we must reformulate the idea of an inner product for this new category of objects.

So, thinking of $\mathbb{C}^n$ as a vector space over $\mathbb{C}$, ...

How do we define an inner product on $\mathbb{C}^n$?

Can <u>not</u> use $\sum_i v_i w_i$!

<u>Ex:</u> $\vec{v} = \begin{pmatrix} 1 \\ i \end{pmatrix}$, $\sum_i v_i v_i = 1 \cdot 1 + i \cdot i = 0$

This defies one of the desirable properties of an inner prod!

<u>Def:</u> The <u>Hermitian dot product</u> on $\mathbb{C}^n$ is

$$\langle \vec{v}, \vec{w} \rangle_H = \sum v_i \overline{w_i} = \vec{v}^T \overline{\vec{w}}$$

<u>Properties:</u> ① $\langle \vec{v}, \vec{w} \rangle_H = \overline{\langle \vec{w}, \vec{v} \rangle_H}$

② $\langle \vec{u} + \vec{v}, \vec{w} \rangle_H = \langle \vec{u}, \vec{w} \rangle_H + \langle \vec{v}, \vec{w} \rangle_H$

③ $\langle c\vec{v}, \vec{w} \rangle_H = c \langle \vec{v}, \vec{w} \rangle_H$

$\langle \vec{v}, c\vec{w} \rangle_H = \overline{c} \langle \vec{v}, \vec{w} \rangle_H$ ⟸ careful!!

④ $\langle \vec{v}, \vec{v} \rangle_H \geq 0$, equality iff $\vec{v} = \vec{0}$

$$\langle \vec{v}, \vec{v} \rangle_H = \sum v_k \overline{v_k} = \sum (a_k + i b_k)(a_k - i b_k)$$

$$= \sum a_k^2 + b_k^2$$

<u>Obs:</u> With the "forgetful function" $\mathbb{C}^n \to \mathbb{R}^{2n}$, the "Hermitian norm" corresponds to the usual "length".

Comment: Any function satisfying ①-④ is called a "Hermitian inner product" on $\mathbb{C}^n$.

Recall that real dot products relate to transposes by:

$$\langle A\vec{v}, \vec{w}\rangle = (A\vec{v})^T \vec{w} = \vec{v}^T A^T \vec{w} = \langle \vec{v}, A^T\vec{w}\rangle$$

So

$$\langle A\vec{v}, \vec{w}\rangle = \langle \vec{v}, A^T\vec{w}\rangle$$

With the Hermitian inner product there is a similar connection to a similar construction.

Def:) The Hermitian transpose of A is

$$A^* = \overline{A}^T$$

Thm:) $\langle A\vec{v}, \vec{w}\rangle_H = \langle \vec{v}, A^*\vec{w}\rangle_H$

Terminology: "Hermitian transpose" = "Hermitian conjugate"
= "conjugate transpose"
= "adjoint"

Recall that transposes relate to symmetry by definition.

We make an analogous definition for Hermitian transpose;

Def: A is <u>Hermitian</u> if $A^* = A$.

(One might think of this as being "Hermitian symmetric".)

So, for symmetric matrices : $\langle A\vec{v}, \vec{w} \rangle = \langle \vec{v}, A\vec{w} \rangle$

and for Hermitian matrices : $\langle A\vec{v}, \vec{w} \rangle_H = \langle \vec{v}, A\vec{w} \rangle_H$

Symmetric Matrices

Note first that real symmetric matrices are Hermitian.

From this we will get some remarkable facts about real symmetric matrices.

Thm:) If A is <u>real</u> and <u>symmetric</u>, then all of its eigenvalues are <u>real</u>, and all of its eigenspaces have <u>real</u> bases.

Pf:) Let λ be an eigenvalue, and $\vec{v}$ a corresponding nonzero eigenvector. Consider the Hermitian inner product:

$$\langle A\vec{v}, \vec{v} \rangle_H$$

(Remember, $\vec{v}$ might be complex...)

Since A is real and symmetric, it is Hermitian.

So

$$\langle A\vec{v}, \vec{v} \rangle_H = \langle \vec{v}, A\vec{v} \rangle_H$$

$$\langle \lambda\vec{v}, \vec{v} \rangle_H = \langle \vec{v}, \lambda\vec{v} \rangle_H$$

$$\lambda \langle \vec{v}, \vec{v} \rangle_H = \bar{\lambda} \langle \vec{v}, \vec{v} \rangle_H$$

$$\lambda = \bar{\lambda}$$

So λ is real.

We find a basis for the eigenspace by finding a basis for $NS(\lambda I - A)$, which is real, so we get a real basis.

Thm: If A is <u>real</u> and <u>symmetric</u>, with eigenvalues $\lambda_1 \neq \lambda_2$ and associated eigenvectors $\vec{v}_1, \vec{v}_2$, then $\vec{v}_1, \vec{v}_2$ are <u>orthogonal</u>.

Pf: Again, A is Hermitian, so

$$\langle A\vec{v}_1, \vec{v}_2 \rangle_H = \langle \vec{v}_1, A\vec{v}_2 \rangle_H$$

$$\langle \lambda_1\vec{v}_1, \vec{v}_2 \rangle_H = \langle \vec{v}_1, \lambda_2\vec{v}_2 \rangle_H$$

$$\lambda_1 \langle \vec{v}_1, \vec{v}_2 \rangle_H = \lambda_2 \langle \vec{v}_1, \vec{v}_2 \rangle_H$$

(because we know λ_2 is real)

$$(\lambda_1 - \lambda_2) \langle \vec{v}_1, \vec{v}_2 \rangle_H = 0$$

$(\neq 0)$ → $(\lambda_1 - \lambda_2)$

$$\langle \vec{v}_1, \vec{v}_2 \rangle_H = 0$$

So $\vec{v}_1, \vec{v}_2$ are orthogonal.

Recall that: A, B <u>similar</u> iff $B = P^{-1}AP$

P represents a change of basis that "turns A into B", or arguably the "way these matrices are similar".

Def: If A, B are similar with $B = P^{-1}AP$, and if P is orthogonal, then we say A, B are <u>orthogonally similar</u>. (and $B = P^T AP$)

Def: If A is diagonalizable with $D = P^{-1}AP$ (so columns of P are a basis of eigenvectors), and if P is orthogonal, then we say A is <u>orthogonally diagonalizable</u>.

Thm: Every real symmetric matrix is orthogonally diagnalizable.

Thm: (Schur) If A has all real eigenvalues, then A is orthogonally similar to an upper triangular matrix.

There is a complex version....

Recall: (columns of P are orthonormal) $\Leftrightarrow$ (P is "orthogonal") $\Leftrightarrow$ ($P^TP=I$)

Def: (columns of P are orthonormal (Hermitian!)) $\Leftrightarrow$ (P is "<u>unitary</u>") $\Leftrightarrow$ ($P^*P=I$)

Like orthogonal matrices, unitary matrices "preserve dot products" (Hermitian!) ;

$$\langle P\vec{v}, P\vec{w} \rangle_H = \langle \vec{v}, \vec{w} \rangle_H$$

Def: If A,B similar with $B=P^{-1}AP$, and if P is unitary, then we say A,B are <u>unitarily similar</u>. (and $B=P^*AP$)

<u>Thm: (Schur)</u> Every square matrix is unitarily similar to an upper triangular matrix.

6.1 – Systems of First Order Linear DE's

We will be interested in systems of DE's of the form:

$$y_1' = a_{11}(x)\, y_1 + \cdots + a_{1n}(x)\, y_n + g_1(x)$$

$$\vdots$$

$$y_n' = a_{n1}(x)\, y_1 + \cdots + a_{nn}(x)\, y_n + g_n(x)$$

If we write $\vec{y} = \begin{pmatrix} y_1 \\ \vdots \\ y_n \end{pmatrix}$, $\vec{g} = \begin{pmatrix} g_1 \\ \vdots \\ g_n \end{pmatrix}$, $A = \begin{pmatrix} a_{11}(x) & \cdots & a_{1n}(x) \\ \vdots & & \vdots \\ a_{n1}(x) & \cdots & a_{nn}(x) \end{pmatrix}$

then this becomes

$$\vec{y}\,'(x) = A\,\vec{y}(x) + \vec{g}$$

Variations ① If $\vec{g}(x) = \vec{0}$, we call this a homogeneous system

② If it is required that $\vec{y}(x_0) = \vec{b}$, this is called an initial value problem.

There is an analogy between the theory of first order linear systems and linear DE's.

Thm:) If $a_{ij}(x)$ and $g_i(x)$ are all continuous on (a,b) containing x_0, then the initial value problem

$$\vec{y}' = A\vec{y} + \vec{g} \quad , \quad \vec{y}(x_0) = \vec{b}$$

has a unique solution on (a,b).

We will not prove this in this course.

Note that if $\vec{y}(x)$ is a solution to

$$\vec{y}' = A\vec{y}$$

then $\vec{y} = \begin{pmatrix} y_1 \\ \vdots \\ y_n \end{pmatrix}$ where each $y_i \in C^1(a,b)$. This is a vector space:

$$(C^1)^n = \left\{ \vec{f} = \begin{pmatrix} f_1 \\ \vdots \\ f_n \end{pmatrix} \middle| \ f_i \in C^1(a,b) \right\}$$

And the set of solutions to the homogeneous system is a subspace, because it is the kernel of the linear operator

$$L(\vec{f}) = \vec{f}' - A\vec{f}$$

(Check that this is linear.)

Thm: The solutions to $\vec{y}' = A\vec{y}$ form a subspace of dimension n.

The proof of this is analogous to the similar result about nth order linear DE's. Roughly, the result again comes from the fact that the space of initial conditions is again n-dimensional.

A basis for this subspace is called a fundamental set of solutions. Such a basis, $\{\vec{Y}_1, \ldots, \vec{Y}_n\}$ can be used to form columns of a matrix called a matrix of fundamental solutions.

$$M = \begin{pmatrix} | & & | \\ \vec{Y}_1 & \cdots & \vec{Y}_n \\ | & & | \end{pmatrix} = \begin{pmatrix} Y_{11} & \cdots & Y_{1n} \\ \vdots & & \vdots \\ Y_{n1} & \cdots & Y_{nn} \end{pmatrix}$$

(Do not confuse these indices, distinguishing solution vectors, with those used to distinguish the variables making up those vectors!)

Ex: Consider the system

$$\vec{y}' = \begin{pmatrix} 1 & -3 \\ -2 & 2 \end{pmatrix} \vec{y}$$

This has a <u>2-dimensional</u> set of solutions.

Note the solutions below are independent:

$$\vec{y}_1 = \begin{pmatrix} -e^{4x} \\ e^{4x} \end{pmatrix}, \quad \vec{y}_2 = \begin{pmatrix} 3e^{-x} \\ 2e^{-x} \end{pmatrix}$$

So, the complete set of solutions is

$$\vec{y} = c_1 \begin{pmatrix} -e^{4x} \\ e^{4x} \end{pmatrix} + c_2 \begin{pmatrix} 3e^{-x} \\ 2e^{-x} \end{pmatrix}$$

and a fundamental set of solutions is

$$\left\{ \begin{pmatrix} -e^{4x} \\ e^{4x} \end{pmatrix}, \begin{pmatrix} 3e^{-x} \\ 2e^{-x} \end{pmatrix} \right\}$$

and the matrix of fundamental solutions is

$$M = \begin{pmatrix} -e^{4x} & 3e^{-x} \\ e^{4x} & 2e^{-x} \end{pmatrix}$$

Note the complete set of solutions is

$$\vec{y} = M\vec{c}, \quad \vec{c} = \begin{pmatrix} c_1 \\ c_2 \end{pmatrix}$$

As with linear DE's, non-homogeneous solutions relate to homogeneous solutions.

Thm:) Say $\{\vec{y}_1, \dots, \vec{y}_n\}$ is a fund. set of sols. for

$$\vec{y}' = A\vec{y}$$

and suppose also that $\vec{y}_p$ is a solution to

$$\vec{y}' = A\vec{y} + \vec{g}$$

Then the complete set of solutions to the non-homogeneous system is

$$\vec{y} = c_1\vec{y}_1 + \dots + c_n\vec{y}_n + \vec{y}_p$$

(Proof is analogous to similar previous thms.)

About the Wronskian ...

Say $v_1, \dots, v_n \in V$; how to tell if l.i. or l.d.?

Suppose you have a l.t. $T: V \to \mathbb{R}^n$. Linearity gives us:

① $\left(\{v_1, \dots, v_n\} \text{ l.d.}\right) \Longrightarrow \left(\{T(v_1), \dots, T(v_n)\} \text{ l.d.}\right)$

and thus

② $\left(\{T(v_1), \dots, T(v_n)\} \text{ l.i.}\right) \Longrightarrow \left(\{v_1, \dots, v_n\} \text{ l.i.}\right)$

But this ↗ is easy to check, because

③ $\left(\{T(v_1), \dots, T(v_n)\} \text{ l.i.}\right) \Longleftrightarrow \det\begin{pmatrix} T(v_1) & \cdots & T(v_n) \end{pmatrix} \neq 0$

So: <u>Any l.t. $T: V \to \mathbb{R}^n$ results in a Wronskian-like construction!</u>

In Chapter 2, we used

$$T(f) = \begin{pmatrix} f \\ f' \\ \vdots \\ f^{[n-1]} \end{pmatrix} \quad \text{and got} \quad W = \det\begin{pmatrix} f_1 & & f_n \\ f_1' & \cdots & f_n' \\ \vdots & & \vdots \\ f_1^{[n-1]} & & f_n^{[n-1]} \end{pmatrix}$$

Here in Chapter 6 we can use instead

$$T\begin{pmatrix} y_1 \\ \vdots \\ y_n \end{pmatrix} = \begin{pmatrix} y_1 \\ \vdots \\ y_n \end{pmatrix}$$

Consider the system

$$\vec{y}' = A\vec{y}$$

Given solutions $\vec{y}_1, \dots, \vec{y}_n$, we define

$$W(x) = \det \begin{pmatrix} | & & | \\ \vec{y}_1 & \cdots & \vec{y}_n \\ | & & | \end{pmatrix}$$

<u>Thm:</u> If $\vec{y}_1, \dots, \vec{y}_n$ are solutions to $\vec{y}' = A\vec{y}$, then either

(1) $W(x)$ is <u>identically zero</u> and $\{\vec{y}_1, \dots, \vec{y}_n\}$ <u>l.d.</u>

<u>or</u>

(2) $W(x)$ is <u>never zero</u> and $\{\vec{y}_1, \dots, \vec{y}_n\}$ l.i.

<u>Ex:</u> Consider the solutions found previously to

$$\vec{y}' = A\vec{y}, \quad A = \begin{pmatrix} 1 & -3 \\ -2 & 2 \end{pmatrix}$$

The Wronskian is

$$W(x) = \det \begin{pmatrix} -e^{4x} & 3e^{-x} \\ e^{4x} & 2e^{-x} \end{pmatrix} = -5e^{3x}$$

This is <u>never zero</u>, so these solutions are l.i..

Constant Coefficient Homogeneous Systems — Diagonalizable

In this section and the next, we will consider how to solve systems

$$\vec{y}\,' = A\vec{y}$$

where A is a constant matrix.

We know that a change of basis can simplify a matrix.

But could this allow a change of basis to simplify a system?

Suppose A is diagonalizable, with

$$D = P^{-1}AP \quad \text{and} \quad A = PDP^{-1}$$

We can then rewrite

$$\vec{y}\,' = A\vec{y}$$

as

$$\vec{y}\,' = PDP^{-1}\vec{y}$$

It is not clear this is better...

... until we instead write

$$P^{-1}\vec{y}' = D P^{-1}\vec{y}$$

and choose to set $\vec{z} = P^{-1}\vec{y}$, giving us

$$\vec{z}' = D\vec{z}$$

This is a <u>diagonal</u> system, which is easy to solve!

$$z_1' = \lambda_1 z_1$$
$$\vdots$$
$$z_n' = \lambda_n z_n$$

(recall the diag. entries of D are the eigenvalues)

(These equations are now independent of each other; we say that the system has been <u>decoupled</u>.)

Recall $P = [I]_{\mathcal{V}}^{\mathcal{S}}$, and $\mathcal{V}$ is a <u>basis of eigenvectors</u>.

$\mathcal{V}$ simplifies a linear transformation ;

$$[T]_{\mathcal{S}}^{\mathcal{S}} = A \quad ☹$$
$$[T]_{\mathcal{V}}^{\mathcal{V}} = D \quad ☺$$

$\mathcal{V}$ simplifies a const. coeff. system

$$\vec{y}' = A\vec{y} \quad ☹$$
$$\vec{z}' = D\vec{z} \quad ☺$$

We can solve the decoupled equations to get

$$z_1 = c_1 e^{\lambda_1 x}, \quad z_2 = c_2 e^{\lambda_2 x}, \quad \ldots, \quad z_n = c_n e^{\lambda_n x}$$

Then

$$\vec{z} = \begin{pmatrix} c_1 e^{\lambda_1 x} \\ \vdots \\ c_n e^{\lambda_n x} \end{pmatrix} = c_1\left(e^{\lambda_1 x}\vec{e}_1\right) + \cdots + c_n\left(e^{\lambda_n x}\vec{e}_n\right)$$

So a fundamental set of solutions to $\vec{z}\,' = D\vec{z}$ is

$$\left\{ e^{\lambda_1 x}\vec{e}_1, \ldots, e^{\lambda_n x}\vec{e}_n \right\}$$

Recall

$$\underbrace{\begin{pmatrix} \lambda_1 & & 0 \\ & \ddots & \\ 0 & & \lambda_n \end{pmatrix}}_{D} = \begin{pmatrix} P^{-1} \end{pmatrix} \begin{pmatrix} A \end{pmatrix} \underbrace{\begin{pmatrix} | & & | \\ \vec{v}_1 & \cdots & \vec{v}_n \\ | & & | \end{pmatrix}}_{P}$$

eigenvalues (the λ_i), eigenvectors (the $\vec{v}_i$)

So when we solve for $\vec{y}$ with $\vec{y} = P\vec{z}$, we get a fundamental set of solutions

$$\left\{ e^{\lambda_1 x} P\vec{e}_1, \ldots, e^{\lambda_n x} P\vec{e}_n \right\}$$

$$= \left\{ e^{\lambda_1 x}\vec{v}_1, \ldots, e^{\lambda_n x}\vec{v}_n \right\}$$

This proves the following theorem.

Thm:) Consider $\vec{y}' = A\vec{y}$, and suppose A is diagonalizable with:

eigenvalues: $\lambda_1, \lambda_2, \dots, \lambda_n$ ⇄ (respectively!)
eigenvectors: $\vec{v}_1, \vec{v}_2, \dots, \vec{v}_n$

Then $\{e^{\lambda_1 x}\vec{v}_1, \dots, e^{\lambda_n x}\vec{v}_n\}$

is a fundamental set of solutions

Ex:) Solve $\vec{y}' = A\vec{y}$, with $A = \begin{pmatrix} 1 & -3 \\ -2 & 2 \end{pmatrix}$. Recall

eigenvalues: $-1, 4$
eigenvectors: $\begin{pmatrix} 3 \\ 2 \end{pmatrix}, \begin{pmatrix} 1 \\ -1 \end{pmatrix}$

Then a f.s.s. is $\left\{ e^{-x}\begin{pmatrix} 3 \\ 2 \end{pmatrix}, e^{4x}\begin{pmatrix} 1 \\ -1 \end{pmatrix} \right\}$

and the general solution is

$$\vec{y} = \begin{pmatrix} 3c_1 e^{-x} + c_2 e^{4x} \\ 2c_1 e^{-x} - c_2 e^{4x} \end{pmatrix}$$

Ex: Solve $\vec{y}' = A\vec{y}$, with $A = \begin{pmatrix} 1 & 1 & 1 \\ 0 & 1 & -1 \\ 0 & 1 & 1 \end{pmatrix}$

eigenvalues: 1, $1+i$, $1-i$

eigenvectors: $\begin{pmatrix} 1 \\ 0 \\ 0 \end{pmatrix}$, $\begin{pmatrix} 1-i \\ i \\ 1 \end{pmatrix}$, $\begin{pmatrix} 1+i \\ -i \\ 1 \end{pmatrix}$

Then a f.s.s. is $\left\{ e^x \begin{pmatrix} 1 \\ 0 \\ 0 \end{pmatrix}, e^{(1+i)x} \begin{pmatrix} 1-i \\ i \\ 1 \end{pmatrix}, e^{(1-i)x} \begin{pmatrix} 1+i \\ -i \\ 1 \end{pmatrix} \right\}$

Observe that these last two solutions are conjugates of each other — so we can replace them with Re, Im components.

$$e^{(1+i)x} \begin{pmatrix} 1-i \\ i \\ 1 \end{pmatrix} = e^x (\cos x + i \sin x) \begin{pmatrix} 1-i \\ i \\ 1 \end{pmatrix}$$

$$= e^x \begin{pmatrix} \cos x + \sin x \\ -\sin x \\ \cos x \end{pmatrix} + i e^x \begin{pmatrix} \sin x - \cos x \\ \cos x \\ \sin x \end{pmatrix}$$

So our <u>real</u> f.s.s. is

$$\left\{ e^x \begin{pmatrix} 1 \\ 0 \\ 0 \end{pmatrix}, e^x \begin{pmatrix} \cos x + \sin x \\ -\sin x \\ \cos x \end{pmatrix}, e^x \begin{pmatrix} \sin x - \cos x \\ \cos x \\ \sin x \end{pmatrix} \right\}$$

Constant Coefficient Homogeneous Systems — Nondiagonalizable

We can try again to decouple...

$$\vec{y}' = A\vec{y}$$

$$\vec{y}' = PJP^{-1}\vec{y}$$

$$P^{-1}\vec{y}' = JP^{-1}\vec{y}$$

$$\vec{z}' = J\vec{z} \qquad (\text{with } \vec{y} = P\vec{z})$$

... but we fail because A is not diagonalizable.

Still, the $\vec{z}$ system is simpler, because J has a lot of zeroes.

Ex:) Solve $\vec{y}' = A\vec{y}$, with

$$A = \begin{pmatrix} -1 & 3 & 5 \\ -4 & 6 & 4 \\ -3 & 2 & 7 \end{pmatrix} = \underbrace{\begin{pmatrix} 1 & 4 & 6 \\ 0 & 2 & 3 \\ 1 & 3 & 5 \end{pmatrix}}_{P} \underbrace{\begin{pmatrix} 4 & 1 & 0 \\ 0 & 4 & 1 \\ 0 & 0 & 4 \end{pmatrix}}_{J} \begin{pmatrix} 1 & 4 & 6 \\ 0 & 2 & 3 \\ 1 & 3 & 5 \end{pmatrix}^{-1}$$

We start with

$$z_1' = 4z_1 + z_2$$

$$z_2' = 4z_2 + z_3$$

$$z_3' = 4z_3$$

Start with z_3, and then "back solve".

3rd eq.: $z_3' = 4z_3 \quad \Rightarrow \quad z_3 = c_3 e^{4x}$

2nd eq.: $z_2' = 4z_2 + c_3 e^{4x}$

Homog. sol.: $z_{2H} = c_2 e^{4x}$

Partic. sol.: Undetermined coeffs, with root!

guess $z_{2p} = Axe^{4x}$

$$Ae^{4x} + 4Axe^{4x} = 4Axe^{4x} + c_3 e^{4x}$$

$$\Rightarrow A = c_3$$

So $z_2 = z_{2H} + z_{2p}$

$$= c_2 e^{4x} + c_3 x e^{4x}$$

1st eq.: $z_1' = 4z_1 + (c_2 e^{4x} + c_3 x e^{4x})$

Homog. sol.: $z_{1H} = c_1 e^{4x}$

Partic. sol.: Undetermined coeffs, with root!

guess $z_{1p} = (Bx + Cx^2)e^{4x}$

$$(B + 2Cx)e^{4x} + 4(Bx + Cx^2)e^{4x} = 4(Bx + Cx^2)e^{4x} + (c_2 e^{4x} + c_3 x e^{4x})$$

$$\Rightarrow B = c_2, \quad C = c_3/2$$

So $z_1 = z_{1H} + z_{1p}$

$$= c_1 e^{4x} + c_2 x e^{4x} + \tfrac{1}{2} c_3 x^2 e^{4x}$$

All together then,

$$\vec{z} = \begin{pmatrix} z_1 \\ z_2 \\ z_3 \end{pmatrix} = \begin{pmatrix} c_1 e^{4x} + c_2 x e^{4x} + \frac{1}{2} c_3 x^2 e^{4x} \\ c_2 e^{4x} + c_3 x e^{4x} \\ c_3 e^{4x} \end{pmatrix}$$

So a f.s.s. for $\vec{z}' = J\vec{z}$ is

$$\left\{ e^{4x}\begin{pmatrix} 1 \\ 0 \\ 0 \end{pmatrix}, \; e^{4x}\begin{pmatrix} x \\ 1 \\ 0 \end{pmatrix}, \; e^{4x}\begin{pmatrix} \frac{1}{2}x^2 \\ x \\ 1 \end{pmatrix} \right\}$$

When we solve for $\vec{y}$ with $\vec{y} = P\vec{z}$, we get

a f.s.s.

$$\left\{ e^{4x}\begin{pmatrix} 1 \\ 0 \\ 1 \end{pmatrix}, \; e^{4x}\begin{pmatrix} 4+x \\ 2 \\ 3+x \end{pmatrix}, \; e^{4x}\begin{pmatrix} 6+4x+\frac{1}{2}x^2 \\ 3+2x \\ 5+3x+\frac{1}{2}x^2 \end{pmatrix} \right\}$$

<u>Question</u>: What if we are not given a Jordan basis?

This is a problem... finding a Jordan basis in general is beyond the scope of this course.

Can use a CAS.

(Matlab : [P,J] = jordan(A))

Easy special case of 2x2 matrices

Note that there is only <u>one</u> Jordan structure for nondiagonalizable 2x2 matrices:

$$J = \begin{pmatrix} \lambda & 1 \\ 0 & \lambda \end{pmatrix} = [T]_{\mathcal{V}}^{\mathcal{V}}$$

where $\mathcal{V} = \{\vec{v}_1, \vec{v}_2\}$, $\vec{v}_1$ is an eigenvector we can find; but suppose we can't find $\vec{v}_2$.

Consider the basis $\mathcal{W} = \{\vec{v}_1, \vec{w}\}$, where $\vec{w} = a\vec{v}_1 + b\vec{v}_2$ is <u>any</u> vector not parallel to $\vec{v}_1$.

We already know

$$A\vec{v}_1 = \lambda\vec{v}_1$$

$$A\vec{v}_2 = 1\vec{v}_1 + \lambda\vec{v}_2$$

Then

$$A\vec{w} = a(\lambda\vec{v}_1) + b(\vec{v}_1 + \lambda\vec{v}_2)$$

$$= b\vec{v}_1 + \lambda\vec{w}$$

So

$$[T]_{\mathcal{W}}^{\mathcal{W}} = \begin{pmatrix} \lambda & b \\ 0 & \lambda \end{pmatrix}$$

This is not necessarily Jordan form... but it is okay for back solving!

Ex:) Solve $\vec{y}' = A\vec{y}$, $A = \begin{pmatrix} 4 & 1 \\ -1 & 6 \end{pmatrix}$

You can confirm that $\begin{pmatrix} 1 \\ 1 \end{pmatrix}$ is the only eigenvector.

Arbitrarily choose $\vec{w} = \begin{pmatrix} -1 \\ 1 \end{pmatrix}$ to be the other basis vector. Then $\mathcal{W} = \left\{ \begin{pmatrix} 1 \\ 1 \end{pmatrix}, \begin{pmatrix} -1 \\ 1 \end{pmatrix} \right\}$, and

$$[T]_{\mathcal{W}}^{\mathcal{W}} = \begin{pmatrix} 1/2 & 1/2 \\ -1/2 & 1/2 \end{pmatrix} \begin{pmatrix} 4 & 1 \\ -1 & 6 \end{pmatrix} \underbrace{\begin{pmatrix} 1 & -1 \\ 1 & 1 \end{pmatrix}}_{P}$$

$$= \begin{pmatrix} 5 & 2 \\ 0 & 5 \end{pmatrix} = K$$

Then $\vec{z}' = K\vec{z}$ can be back solved, and $\vec{y}_i = P\vec{z}_i$ form the f.s.s. for the original system.

$$\vec{y}' = A\vec{y} \quad K = P^{-1}AP$$

$$\vec{y}' = (PKP^{-1})\vec{y}$$

$$(P^{-1}\vec{y})' = K(P^{-1}\vec{y}) \quad \vec{z} = P^{-1}\vec{y}, \text{ or } \vec{y} = P\vec{z}$$

$$\vec{z}' = K\vec{z}$$

Alternative method

We define the "matrix exponential" by

$$e^M = I + M + \frac{M^2}{2!} + \frac{M^3}{3!} + \dots$$

(analogous to how complex exponentials are motivated!)

Note $\left(e^{xA}\right)' = \left(I + (xA) + \frac{(xA)^2}{2!} + \frac{(xA)^3}{3!} + \dots\right)'$

$$= \left(\quad A + xA^2 + \frac{x^2}{2!}A^3 + \dots\right)$$

$$= A\left(I + (xA) + \frac{(xA)^2}{2!} + \dots\right)$$

$$= Ae^{xA}$$

Using this we can directly observe that the solution to the I.V.P.

$$\vec{Y}' = A\vec{Y} \quad , \vec{Y}(0) = \vec{Y}_0$$

is $\vec{Y} = e^{xA}\vec{Y}_0$! (note also that $e^0 = I$.)

So every solution to $\vec{Y}' = A\vec{Y}$ must take this form.

Thm:) If $\{\vec{v}_1, \ldots, \vec{v}_n\}$ is any basis for $\mathbb{R}^n$, then

$$\{e^{xA}\vec{v}_1, \ldots, e^{xA}\vec{v}_n\}$$ is a fss to $\vec{Y}' = A\vec{Y}$.

Pf:) (Show that $W(0)$ is nonzero!)

But how do we evaluate e^{xA}?

Thm:) $$e^{xA}\vec{v} = e^{\lambda x}\left(I + x(A-\lambda I) + \frac{x^2}{2}(A-\lambda I)^2 + \ldots\right)\vec{v}$$

Cor.:) If $\vec{v}$ is an eigenvector with eigenvalue λ, then

$$e^{xA}\vec{v} = e^{\lambda x}\vec{v}$$

(NB, if you apply the basis theorem above to an eigenbasis, you reproduce exactly the fss we have already found for a diagonalizable system!)

For the nondiagonalizable case, we need to understand what $(A-\lambda I)$ does to a Jordan basis vector – and then choose a Jordan basis.

Let's look at a $\underline{\text{single}}$ Jordan block, and the corresponding Jordan basis vectors.

$$J = \begin{bmatrix} \ddots & & & & & \\ & \lambda & 1 & & & \\ & & \lambda & 1 & & \\ & & & \lambda & \ddots & \\ & & & & \ddots & 1 \\ & & & & & \lambda \\ & & & & & & \ddots \end{bmatrix} = [T]_{\mathcal{V}}^{\mathcal{V}}$$

$\mathcal{V}$ ← Jordan basis

Columns: $\vec{w}_1 \; \vec{w}_2 \; \vec{w}_3 \; \cdots \; \vec{w}_k$

For $\vec{w}_1$: $\quad T(\vec{w}_1) = \lambda \vec{w}_1$

For $\vec{w}_2, \ldots, \vec{w}_k$: $\quad T(\vec{w}_i) = 1\,\vec{w}_{i-1} + \lambda \vec{w}_i$

Interpreting w.r.t. $\mathcal{S}$, we get

$$\begin{aligned} A\vec{w}_1 &= \lambda \vec{w}_1 \\ A\vec{w}_i &= \vec{w}_{i-1} + \lambda \vec{w}_i \end{aligned} \quad \Longrightarrow \quad \begin{aligned} (A-\lambda I)\vec{w}_1 &= 0 \\ (A-\lambda I)\vec{w}_i &= \vec{w}_{i-1} \end{aligned}$$

So,

$$\vec{0} \longleftarrow \vec{w}_1 \xleftarrow{(A-\lambda I)} \vec{w}_2 \xleftarrow{(A-\lambda I)} \cdots \xleftarrow{(A-\lambda I)} \vec{w}_{k-1} \xleftarrow{(A-\lambda I)} \vec{w}_k$$

So our previous equation

$$e^{xA}\vec{v} = e^{\lambda x}\left(I + x(A-\lambda I) + \frac{x^2}{2}(A-\lambda I)^2 + \dots\right)\vec{v}$$

in the case of a Jordan basis vector $\vec{w}_i$ is

$$e^{xA}w_i = e^{\lambda x}\left(\vec{w}_i + x\vec{w}_{i-1} + \frac{x^2}{2}\vec{w}_{i-2} + \dots + \frac{x^{i-1}}{(i-1)!}\vec{w}_1\right)$$

Ex: Solve $\vec{y}\,' = A\vec{y}$, with

$$A = \begin{pmatrix} -1 & 3 & 5 \\ -4 & 6 & 4 \\ -3 & 2 & 7 \end{pmatrix} = \begin{pmatrix} 1 & 4 & 6 \\ 0 & 2 & 3 \\ 1 & 3 & 5 \end{pmatrix}\begin{pmatrix} 4 & 1 & 0 \\ 0 & 4 & 1 \\ 0 & 0 & 4 \end{pmatrix}\begin{pmatrix} 1 & 4 & 6 \\ 0 & 2 & 3 \\ 1 & 3 & 5 \end{pmatrix}^{-1}$$

The Jordan basis given in the above equation is

$$\vec{w}_1 = \begin{pmatrix} 1 \\ 0 \\ 1 \end{pmatrix}, \quad \vec{w}_2 = \begin{pmatrix} 4 \\ 2 \\ 3 \end{pmatrix}, \quad \vec{w}_3 = \begin{pmatrix} 6 \\ 3 \\ 5 \end{pmatrix}$$

So a f.s.s. is

$$\left\{e^{xA}\vec{w}_1,\ e^{xA}\vec{w}_2,\ e^{xA}\vec{w}_3\right\}$$

$$= \left\{e^{4x}(\vec{w}_1),\ e^{4x}(\vec{w}_2 + x\vec{w}_1),\ e^{4x}\left(\vec{w}_3 + x\vec{w}_2 + \tfrac{1}{2}x^2\vec{w}_1\right)\right\}$$

$$= \left\{e^{4x}\begin{pmatrix} 1 \\ 0 \\ 1 \end{pmatrix},\ e^{4x}\begin{pmatrix} 4+x \\ 2 \\ 3+x \end{pmatrix},\ e^{4x}\begin{pmatrix} 6+4x+\frac{1}{2}x^2 \\ 3+2x \\ 5+3x+\frac{1}{2}x^2 \end{pmatrix}\right\}$$

Remember, the basis vectors for each Jordan block are grouped separately

Ex: Suppose

$$J = \begin{pmatrix} 3 & 1 & & & \\ & 3 & & & \\ & & 4 & 1 & \\ & & & 4 & \\ & & & & 4 \end{pmatrix} = [T]_{\mathcal{V}}^{\mathcal{V}}, \quad \mathcal{V} = \{\vec{v}_1, \vec{v}_2, \vec{v}_3, \vec{v}_4, \vec{v}_5\}$$

Then

$$0 \xleftarrow{(A-3I)} \vec{v}_1 \xleftarrow{(A-3I)} \vec{v}_2$$

$$0 \xleftarrow{(A-4I)} \vec{v}_3 \xleftarrow{(A-4I)} \vec{v}_4$$

$$0 \xleftarrow{A-4I} \vec{v}_5$$

So, using this Jordan basis, we get the f.s.s.

$$\left\{ e^{3x}(\vec{v}_1),\ e^{3x}(\vec{v}_2 + x\vec{v}_1),\ e^{4x}(\vec{v}_3),\ e^{4x}(\vec{v}_4 + x\vec{v}_3),\ e^{4x}(\vec{v}_5) \right\}$$

6.4 – Nonhomogeneous Linear Systems

Recall that the general solution to

$$\vec{y}' = A\vec{y} + \vec{g}$$

is obtained by finding <u>any</u> particular solution, and then adding the complete homogeneous solutions.

So, how do we find the particular solution?

Let's consider first the single variable case.

$$y' = a(x)\,y + g(x)$$

The homogeneous equation and solution are

$$y' = ay \quad \Rightarrow \quad y = k\,e^{\int a\,dx}$$

So the fundamental solution is $m(x) = e^{\int a(x)\,dx}$

One might speculate that the particular solution could look similar. We could try:

$$y' = ay + g \quad \overset{?}{\Leftarrow} \quad y_p = m(x)\,v(x)$$

and then try to solve for this factor $v(x)$ ...

Plugging in to the equation and doing some algebra, we get

$$(mv)' = amv + g$$

$$m'v + mv' = amv + g$$

(These cancel because $m' = am$)

$$mv' = g$$

$$v = \int m^{-1} g \, dx$$

$$y_p = mv = m \int m^{-1} g \, dx$$

One can then plug this in to the equation to confirm that it actually is a solution.

(See 4.4, variation of parameters)

(Note this gives us the same solution as the method of integrating factors in this case.)

We can do the same in the case of systems.

For the homogeneous system

$$\vec{y}' = A\vec{y}$$

we know how to find a fundamental set of solutions.

Let M be the solution matrix

$$M = \begin{pmatrix} | & & | \\ \vec{y}_1 & \cdots & \vec{y}_n \\ | & & | \end{pmatrix}$$

We might speculate that the particular solution again will look similar to the $\vec{y}$'s ...

$$\vec{y}' = A\vec{y} + \vec{g} \quad \overset{?}{\leftarrow} \quad \vec{y}_p = M\vec{v}$$

Again, we can plug in and solve.

$$(M\vec{v})' = AM\vec{v} + \vec{g}$$

$$M'\vec{v} + M\vec{v}' = AM\vec{v} + \vec{g}$$

(These cancel because $M' = AM$, because every column of M is a solution to $\vec{y}' = A\vec{y}$.)

$$M\vec{v}' = g$$

$$\vec{v} = \int M^{-1}\vec{g}\,dx$$

$$\vec{y}_p = M\vec{v} = \boxed{M\int M^{-1}\vec{g}\,dx}$$

Ex:) Find a particular solution to

$$\vec{y}' = \begin{pmatrix} 1 & -3 \\ -2 & 2 \end{pmatrix} \vec{y} + \begin{pmatrix} x \\ 3x \end{pmatrix}$$

Recall that the matrix of solutions is

$$M = \begin{pmatrix} -e^{4x} & 3e^{-x} \\ e^{4x} & 2e^{-x} \end{pmatrix}$$

Its inverse is

$$M^{-1} = \begin{pmatrix} 2e^{-x} & -3e^{-x} \\ -e^{4x} & -e^{4x} \end{pmatrix} \Big/ (-5e^{3x})$$

$$= \begin{pmatrix} \frac{-2}{5} e^{-4x} & \frac{3}{5} e^{-4x} \\ \frac{1}{5} e^{x} & \frac{1}{5} e^{x} \end{pmatrix}$$

Then $\quad M^{-1} g = \begin{pmatrix} \frac{7}{5} x e^{-4x} \\ \frac{4}{5} x e^{x} \end{pmatrix}$

$$\int M^{-1} g \, dx = \begin{pmatrix} \frac{-7}{20} x e^{-4x} - \frac{7}{80} e^{-4x} \\ \frac{4}{5} x e^{x} - \frac{4}{5} e^{x} \end{pmatrix}$$

$$\vec{y}_p = M \int M^{-1} g \, dx = \begin{pmatrix} \left(\frac{7}{20} x + \frac{7}{80}\right) + \left(\frac{12}{5} x - \frac{12}{5}\right) \\ \left(\frac{-7}{20} x - \frac{7}{80}\right) + \left(\frac{8}{5} x - \frac{8}{5}\right) \end{pmatrix} = \begin{pmatrix} \frac{11}{4} x - \frac{37}{16} \\ \frac{5}{4} x - \frac{27}{16} \end{pmatrix}$$

Ex: Find a particular solution to

$$\vec{y}\,' = \begin{pmatrix} 1 & -3 \\ -2 & 2 \end{pmatrix} \vec{y} + \begin{pmatrix} \cos x \\ 0 \end{pmatrix}$$

We can use complex solutions to solve this real problem. We consider

$$\vec{z}\,' = \begin{pmatrix} 1 & -3 \\ -2 & 2 \end{pmatrix} \vec{z} + \begin{pmatrix} e^{ix} \\ 0 \end{pmatrix}$$

M, M^{-1} are as in the previous example.

$$M^{-1}\begin{pmatrix} e^{ix} \\ 0 \end{pmatrix} = \begin{pmatrix} -\frac{2}{5} e^{(-4+i)x} \\ \frac{1}{5} e^{(1+i)x} \end{pmatrix}$$

$$\int M^{-1}\begin{pmatrix} e^{ix} \\ 0 \end{pmatrix} dx = \begin{pmatrix} -\frac{2}{5} \frac{1}{-4+i} e^{(-4+i)x} \\ \frac{1}{5} \frac{1}{1+i} e^{(1+i)x} \end{pmatrix}$$

$$= \begin{pmatrix} \frac{8+2i}{85} e^{(-4+i)x} \\ \frac{1-i}{10} e^{(1+i)x} \end{pmatrix}$$

$$\vec{z}_p = M\int M^{-1}\begin{pmatrix} e^{ix} \\ 0 \end{pmatrix} dx = \begin{pmatrix} -\frac{8+2i}{85} e^{ix} + \frac{3-3i}{10} e^{ix} \\ \frac{8+2i}{85} e^{ix} + \frac{2-2i}{10} e^{ix} \end{pmatrix}$$

$$= \begin{pmatrix} \left(\frac{7}{34} - \frac{11}{34} i\right)(\cos x + i \sin x) \\ \left(\frac{5}{17} - \frac{3}{17} i\right)(\cos x + i \sin x) \end{pmatrix}$$

Since $\begin{pmatrix} \cos x \\ 0 \end{pmatrix} = \text{Re}\begin{pmatrix} e^{ix} \\ 0 \end{pmatrix}$ we have $\vec{y}_p = \text{Re}(\vec{z}_p)$:

$$\vec{y}_p = \begin{pmatrix} \frac{7}{34}\cos x + \frac{11}{34}\sin x \\ \frac{5}{17}\cos x + \frac{3}{17}\sin x \end{pmatrix}$$

The first example makes clear an awkward point about this technique ... that is, we used complicated algebra in a problem where both the question and the answer are fairly simple.

Instead then, we can try to guess the form of the solution, and use algebra to work out the details. (This is the idea behind "undetermined coefficients" in 4.3.)

Ex:) Find a particular solution to

$$\vec{y}\,' = \begin{pmatrix} 1 & -3 \\ -2 & 2 \end{pmatrix} \vec{y} + \begin{pmatrix} x \\ 3x \end{pmatrix}$$

We guess $\vec{y}_p = \begin{pmatrix} a_1 \\ a_2 \end{pmatrix} + \begin{pmatrix} b_1 \\ b_2 \end{pmatrix} x = \vec{a} + \vec{b}x$

Noting $\vec{y}_p\,' = \vec{b}$, the equation becomes

$$\vec{b} = A(\vec{a} + \vec{b}x) + \begin{pmatrix} 1 \\ 3 \end{pmatrix} x$$

Grouping like powers of x, we get

$$A\vec{b} = \begin{pmatrix} -1 \\ -3 \end{pmatrix} \quad \text{and} \quad A\vec{a} = \vec{b}$$

Since $A^{-1} = \begin{pmatrix} -1/2 & -3/4 \\ -1/2 & -1/4 \end{pmatrix}$ we have

$$\vec{b} = A^{-1} \begin{pmatrix} -1 \\ -3 \end{pmatrix} = \begin{pmatrix} 11/4 \\ 5/4 \end{pmatrix}$$

$$\vec{a} = A^{-1} \begin{pmatrix} 11/4 \\ 5/4 \end{pmatrix} = \begin{pmatrix} -37/16 \\ -27/16 \end{pmatrix}$$

So we get the same particular solution as previous:

$$\vec{y}_p = \begin{pmatrix} -37/16 + 11/4\, x \\ -27/16 + 5/4\, x \end{pmatrix}$$

Ex:) Find a particular solution to

$$\vec{y}' = \begin{pmatrix} 1 & -3 \\ -2 & 2 \end{pmatrix} \vec{y} + e^{rx} \vec{v}$$

Try $\vec{y}_p = e^{rx} \vec{a}$

$$r e^{rx} \vec{a} = A e^{rx} \vec{a} + e^{rx} \vec{v}$$

$$(A - rI) \vec{a} = -\vec{v}$$

You can solve for $\vec{a}$ and get a solution, ...
unless $\det(A - rI) = 0$ ... that is, unless
r is an eigenvalue of A.

Ex:) Find a particular solution to

$$\vec{y}' = \begin{pmatrix} 1 & -3 \\ -2 & 2 \end{pmatrix} \vec{y} + e^{4x} \begin{pmatrix} 1 \\ 2 \end{pmatrix}$$

Note, 4 is an eigenvalue for A.

Try $\vec{y}_p = e^{4x} (\vec{a} + \vec{b} x)$

The equation becomes

$$4\cancel{e^{4x}}(\vec{a}+\vec{b}x)+\cancel{e^{4x}}(\vec{b}) = \cancel{e^{4x}}\,A(\vec{a}+\vec{b}x)+\cancel{e^{4x}}\vec{v}$$

$$x(4\vec{b}-A\vec{b})+(4\vec{a}+\vec{b}-A\vec{a}-\vec{v})=0$$

$$(A-4I)\vec{b}=\vec{0} \quad \underline{\underline{\text{and}}} \quad (A-4I)\vec{a}=\vec{b}-\vec{v}$$

Since 4 is an eigenvalue with eigenvector $\begin{pmatrix}-1\\1\end{pmatrix}$, we conclude from the first equation that

$$\vec{b}=k\begin{pmatrix}-1\\1\end{pmatrix}$$

The second equation then becomes

$$(A-4I)\vec{a}=k\begin{pmatrix}-1\\1\end{pmatrix}-\vec{v}$$

$$\begin{pmatrix}-3 & -3\\-2 & -2\end{pmatrix}\vec{a}=\begin{pmatrix}-k-1\\k-2\end{pmatrix}$$

To have a solution we must have

$$\frac{-k-1}{k-2}=\frac{3}{2} \Rightarrow -2k-2=3k-6$$

$$\Rightarrow 4=5k$$

$$\Rightarrow k=\frac{4}{5}$$

The second equation then becomes

$$\begin{pmatrix} -3 & -3 \\ -2 & -2 \end{pmatrix} \begin{pmatrix} a_1 \\ a_2 \end{pmatrix} = \begin{pmatrix} -9/5 \\ -6/5 \end{pmatrix}$$

So $\quad a_1 + a_2 = 3/5$

and we can choose

$$\vec{a} = \begin{pmatrix} 3/5 \\ 0 \end{pmatrix}$$

Our solution then is

$$\vec{y}_p = e^{4x}\left(\vec{a} + \vec{b}x\right)$$

$$= e^{4x}\begin{pmatrix} 3/5 - 4/5\,x \\ 0 + 4/5\,x \end{pmatrix}$$

Converting Higher Order Equations and Systems

Consider the equation

$$a_3 y''' + a_2 y'' + a_1 y' + a_0 y = g$$

Everything on the left is at most a first derivative...

... of y, y', or y''.

We can represent this by writing

$$y = u_0$$

$$y' = u_1 \qquad = u_0'$$

$$y'' = u_2 \qquad = u_1'$$

and then

$$y''' = -\frac{a_0}{a_3} u_0 - \frac{a_1}{a_3} u_1 - \frac{a_2}{a_3} u_2 - \frac{g}{a_3} \qquad = u_2'$$

This is then a first order system in u_0, u_1, u_2.

$$u_0' = 0\, u_0 + 1\, u_1 + 0\, u_2 + 0$$

$$u_1' = 0\, u_0 + 0\, u_1 + 1\, u_2 + 0$$

$$u_2' = -\frac{a_0}{a_3} u_0 - \frac{a_1}{a_3} u_1 - \frac{a_2}{a_3} u_2 - \frac{g}{a_3}$$

Not clear yet how this is useful...

Thm:) If you convert a CCLDE to a system as above

$$\underbrace{1y^{[n]}+\dots+a_0y=0}_{\substack{\text{char. poly.}\\ =1\lambda^n+\dots+a_0=p_1(\lambda)}} \rightsquigarrow \underbrace{\vec{u}'=A\vec{u}}_{\substack{\text{char. poly.}\\ =\det(A-\lambda I)=p_2(\lambda)}}$$

then $p_1 = p_2$.

Ex:)

$$\underbrace{y''+5y'+6y=0}_{p_1(\lambda)=\lambda^2+5\lambda+6} \rightsquigarrow \underbrace{\vec{u}'=\begin{pmatrix}0 & 1\\ -6 & -5\end{pmatrix}\vec{u}}_{p_2(\lambda)=\det\begin{pmatrix}-\lambda & 1\\ -6 & -5-\lambda\end{pmatrix}}$$

$$= \lambda^2+5\lambda+6$$

At a glance this is surprising. But...

CCLDE	$\rightsquigarrow$	system
sols	$(y=u_0)$ - - - - - -	sols
↑		↑
roots of $p_1(\lambda)$		roots of $p_2(\lambda)$
↑		↑
factors of $p_1(\lambda)$		factors of $p_2(\lambda)$
↑		↑
$p_1(\lambda)$		$p_2(\lambda)$

You can use this same idea to convert higher order systems to first order systems.

Ex:) Consider the system

$$y_1'' = 3y_1 + 2y_1' - y_2 + y_2'$$
$$y_2'' = 2y_1 - 4y_1' + 2y_2 - y_2'$$

We can choose

$$\begin{aligned} y_1 &= u_1 \\ y_1' &= u_2 \\ y_2 &= u_3 \\ y_2' &= u_4 \end{aligned}$$

The system then becomes

$$\begin{aligned} u_1' &= 0u_1 + 1u_2 + 0u_3 + 0u_4 \\ u_2' &= 3u_1 + 2u_2 - 1u_3 + 1u_4 \\ u_3' &= 0u_1 + 0u_2 + 0u_3 + 1u_4 \\ u_4' &= 2u_1 - 4u_2 + 2u_3 - 1u_4 \end{aligned}$$

or $\vec{u}' = A\vec{u}$ with

$$A = \begin{pmatrix} 0 & 1 & 0 & 0 \\ 3 & 2 & -1 & 1 \\ 0 & 0 & 0 & 1 \\ 2 & -4 & 2 & -1 \end{pmatrix}$$

6.6 — Applications Involving Systems

Ex i) Consider a mass on a spring, with another spring and mass hanging from it. How can we describe the motion?

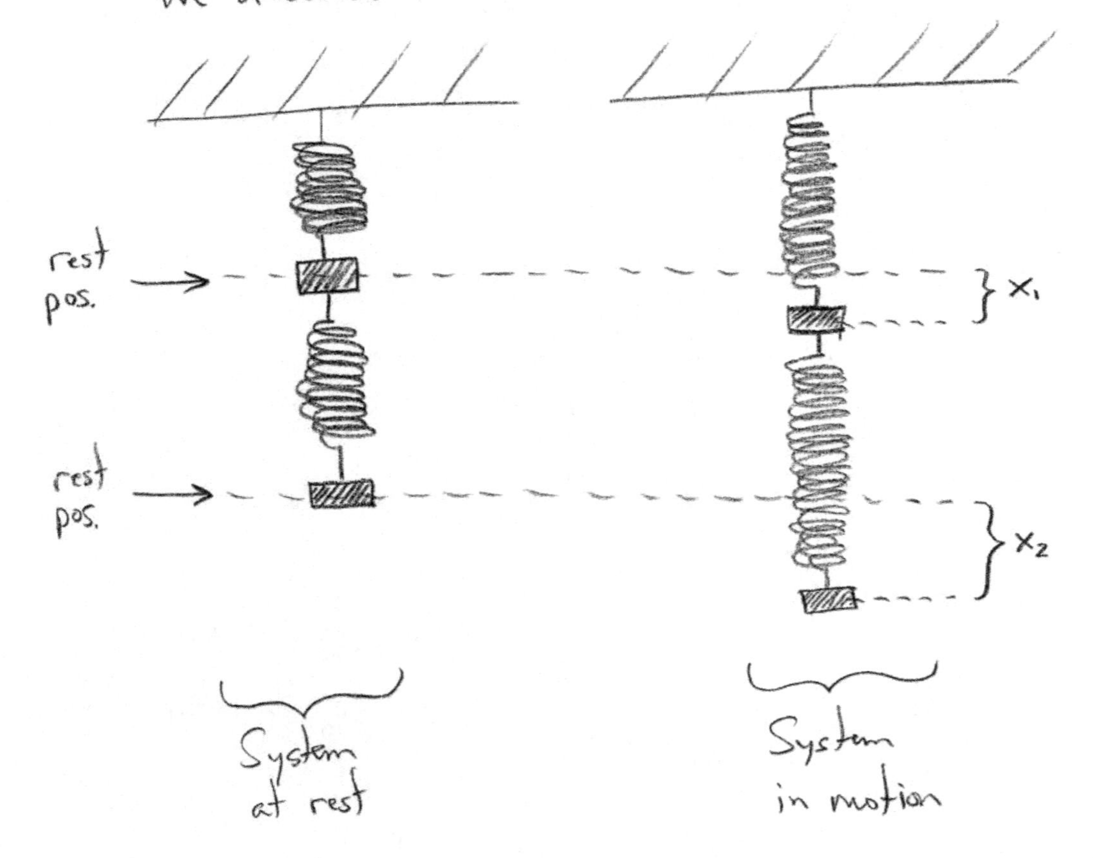

The forces on these masses again come from acceleration, friction, springs, and external.

The system is

$$m_1 x_1'' = -k_1 x_1 + k_2 (x_2 - x_1) - f_1 x_1' + h_1(t)$$

$$m_2 x_2'' = -k_2 (x_2 - x_1) - f_2 x_2' + h_2(t)$$

$$\underbrace{\qquad}_{\text{accel.}} \quad \underbrace{\qquad\qquad\qquad}_{\text{spring forces}} \quad \underbrace{\qquad}_{\text{friction}} \quad \underbrace{\qquad}_{\text{external}}$$

or

$$x_1'' = \left(\frac{-k_1 - k_2}{m_1}\right) x_1 + \left(\frac{-f_1}{m_1}\right) x_1' + \left(\frac{k_2}{m_1}\right) x_2 + (0) x_2' + \frac{h_1}{m_1}$$

$$x_2'' = \left(\frac{k_2}{m_2}\right) x_1 + (0) x_1' + \left(-\frac{k_2}{m_2}\right) x_2 - \left(\frac{f_2}{m_2}\right) x_2' + \frac{h_2}{m_2}$$

Writing

$$x_1 = u_1$$
$$x_1' = u_2$$
$$x_2 = u_3$$
$$x_2' = u_4$$

this becomes

$$u_1' = 0\, u_1 + 1\, u_2 + 0\, u_3 + 0\, u_4 + 0$$

$$u_2' = \left(\frac{-k_1 - k_2}{m_1}\right) u_1 + \left(\frac{-f_1}{m_1}\right) u_2 + \left(\frac{k_2}{m_1}\right) u_3 + 0\, u_4 + \frac{h_1}{m_1}$$

$$u_3' = 0\, u_1 + 0\, u_2 + 0\, u_3 + 1\, u_4 + 0$$

$$u_4' = \left(\frac{k_2}{m_2}\right) u_1 + 0\, u_2 + \left(\frac{-k_2}{m_2}\right) u_3 + \left(\frac{-f_2}{m_1}\right) u_4 + \frac{h_2}{m_2}$$

We can write this in matrix form as

$$\vec{u}' = \begin{pmatrix} 0 & 1 & 0 & 0 \\ \left(\frac{-k_1-k_2}{m_1}\right) & \left(-\frac{f_1}{m_1}\right) & \left(\frac{k_2}{m_1}\right) & 0 \\ 0 & 0 & 0 & 1 \\ \left(\frac{k_2}{m_2}\right) & 0 & \left(\frac{-k_2}{m_2}\right) & \left(\frac{-f_2}{m_2}\right) \end{pmatrix} \vec{u} + \begin{pmatrix} 0 \\ h_1/m_1 \\ 0 \\ h_2/m_1 \end{pmatrix}$$

Ex: If there is no friction and no external forcing, we can solve the second order system <u>without converting</u>.

$$m_1 x_1'' = -k_1 x_1 + k_2 (x_2 - x_1)$$

$$\underbrace{m_2 x_2''}_{\text{accel.}} = \underbrace{-k_2 (x_2 - x_1)}_{\text{spring forces}}$$

$$\vec{x}'' = \begin{pmatrix} \frac{-k_1-k_2}{m_1} & \frac{k_2}{m_1} \\ \frac{k_2}{m_2} & -\frac{k_2}{m_2} \end{pmatrix} \vec{x}$$

This is homogeneous, and we might again hope to find exponential solutions of the form

$$\vec{x} = e^{rt} \vec{v}$$

The equation becomes

$$r^2 e^{rt} \vec{v} = A e^{rt} \vec{v}$$

$$r^2 \vec{v} = A\vec{v}$$

This is an eigenvalue equation, with eigenvalue $\lambda = r^2$ and eigenvector $\vec{v}$.

That is — if we have an eigenvector $\vec{v}$ and choose $r = \pm\sqrt{\lambda}$, then

$$\vec{y} = e^{rt}\vec{v}$$

is a solution.

With some algebra it can be shown that for this particular matrix, λ is always negative. So for any eigenvalue, we choose

$$r = \pm\sqrt{\lambda} = \pm i\sqrt{|\lambda|} = \pm i\omega$$

If we have two eigenvalues λ_1, λ_2 with eigenvectors $\vec{v}_1, \vec{v}_2$, we then get solutions

$$\left\{ e^{i\omega_1 t}\vec{v}_1 ,\ e^{-i\omega_1 t}\vec{v}_1 ,\ e^{i\omega_2 t}\vec{v}_2 ,\ e^{-i\omega_2 t}\vec{v}_2 \right\}$$

Looking at real and imaginary parts, we can choose a different (real) basis:

$$\left\{ \vec{v}_1 \cos(\omega_1 t),\ \vec{v}_1 \sin(\omega_1 t),\ \vec{v}_2 \cos(\omega_2 t),\ \vec{v}_2 \sin(\omega_2 t) \right\}$$

Made in the USA
Columbia, SC
08 January 2020